D1465730

Library

3 9002 10039529 3

TOWARDS THE SEMANTIC WEB

TOWARDS THE SEMANTIC WEB
Ontology-driven Knowledge Management

Edited by

Dr John Davies
British Telecommunications plc

Professor Dieter Fensel
University of Innsbruck, Austria

and Professor Frank van Harmelen
Vrije Universiteit, Amsterdam, Netherlands

JOHN WILEY & SONS, LTD

Copyright © 2003   John Wiley & Sons Ltd,
The Atrium, Southern Gate, Chichester,
West Sussex PO19 8SQ, England

Telephone (+44) 1243 779777

Email (for orders and customer service enquiries): cs-books@wiley.co.uk
Visit our Home Page on www.wileyeurope.com or www.wiley.com

All Rights Reserved. No part of this publication may be reproduced, stored in a retrieval system or transmitted in any form or by any means, electronic, mechanical, photocopying, recording, scanning or otherwise, except under the terms of the Copyright, Designs and Patents Act 1988 or under the terms of a licence issued by the Copyright Licensing Agency Ltd, 90 Tottenham Court Road, London W1T 4LP, UK, without the permission in writing of the Publisher. Requests to the Publisher should be addressed to the Permissions Department, John Wiley & Sons Ltd, The Atrium, Southern Gate, Chichester, West Sussex PO19 8SQ, England, or emailed to permreq@wiley.co.uk, or faxed to (+44) 1243 770571.

This publication is designed to provide accurate and authoritative information in regard to the subject matter covered. It is sold on the understanding that the Publisher is not engaged in rendering professional services. If professional advice or other expert assistance is required, the services of a competent professional should be sought.

Other Wiley Editorial Offices

John Wiley & Sons Inc.,
111 River Street, Hoboken, NJ 07030, USA

Jossey-Bass, 989 Market Street, San Francisco,
CA 94103–1741, USA

Wiley-VCH Verlag GmbH,
Boschstr. 12, D–69469 Weinheim, Germany

John Wiley & Sons Australia Ltd, 33 Park Road,
Milton, Queensland 4064, Australia

John Wiley & Sons (Asia) Pte Ltd, 2 Clementi Loop 02–01,
Jin Xing Distripark, Singapore 129809

John Wiley & Sons Canada Ltd, 22 Worcester Road,
Etobicoke, Ontario, Canada M9W 1L1

Library of Congress Cataloging-in-Publication Data

Towards the semantic web : ontology-driven knowledge management / edited by John Davies, Dieter Fensel, and Frank van Harmelen.
    p. cm.
   Includes bibliographical references and index.
   ISBN 0-470-84867-7 (alk. paper)
    1. Semantic web. 2. Ontology. 3. Knowledge acquisition (Expert systems) I. Davies, John. II. Fensel, Dieter. III. Van Harmelen, Frank.

   TK5105.88815.T68 2002
   006.3'3–dc21

                         2002033103

British Library Cataloguing in Publication Data

A catalogue record for this book is available from the British Library

ISBN 0470 84867 7

Typeset in 10/12pt Times by Deerpark Publishing Services Ltd, Shannon, Ireland.

This book is printed on acid-free paper responsibly manufactured from sustainable forestry in which at least two trees are planted for each one used for paper production.

Limerick Institute of Technology
Institiúid Teicneolaíochta Luimnigh
Library and Information Resource Centre

Class No.: _025.04 DAV_

ACC. No.: _31753_

Date: _04|06|2010_

Contents

13 Field Experimenting with Semantic Web Tools in a Virtual Organization 219
Victor Iosif, Peter Mika, Rikard Larsson and Hans Akkermans

14 A Future Perspective: Exploiting Peer-to-Peer and the Semantic Web for Knowledge Management 245
Dieter Fensel, Steffen Staab, Rudi Studer, Frank van Harmelen and John Davies

Foreword

Knowledge is Power Again!
J. Hendler, University of Maryland

More than 30 years ago, ACM Turing Award winner, Ed Feigenbaum, heralded a revolution in business computing under the banner 'knowledge is power'. With this slogan, Feigenbaum brought domain-specific expert systems to the attention of the computing world. Now deployed in shrink-wrapped tax preparation programs, embedded in one of the world's best selling software products, and estimated to be in use by over two-thirds of Fortune 500 companies, the expert system gains its power by the use of the specific knowledge of a domain that is encoded in its rules – be it rules about tax laws, rules about the spelling of words, or the specific business rules dictating how your market sector operates. In all these systems, this special-purpose knowledge is where the power is derived.

In the past decade, however, a new agenda has been evolving as part of research in what is now known as the Semantic Web. This approach might also be called 'knowledge is power,' but with a significantly different metaphor. Where Feigenbaum envisioned power akin to the power of a sledgehammer, the new paradigm makes knowledge akin to the power flowing through the electrical grid. Rather than the centralized power coming from carefully engineered knowledge bases aimed at specific applications, the new power flows through the routers of the Internet, as electricity flows through the wires in your wall. Knowledge, in this view, becomes as distributed, dynamic and ubiquitous as the power flowing into the lamp by which you are reading these words.

The Semantic Web vision, per se, is rightly attributed to Tim Berners-Lee, inventor of the web and coiner of the term 'Semantic Web,' but he was not the first or only one to realize the strength of the new knowledge is power metaphor. A small group of researchers, branching out from the traditional confines of knowledge representation in Artificial Intelligence, were talking about 'knowledge servers,' 'semantic engines,' 'ontology management systems,' and other approaches to ubiquitous knowledge before the web even came into being. However, with the expanding impact of Berners-Lee's World Wide Web, the deployment vehicle for this ubiquitous knowledge became clear, and these Artificial Intelligence technologies, brought to the web, now provide the knowledge technologies capable of powering the Semantic Web.

The power of the semantic web, therefore, comes from the coupling of the knowledge technologies developed by the AI world with the power grid being developed by the Web developers. Sitting on top of web-embedded languages like the Resource Description Framework (RDF) and the Extensible Markup Language (XML), the new Semantic Web languages bring powerful AI concepts into contact with the Web infrastructure that has changed the world. The Web, reaching into virtually every computer around the world, can now carry the knowledge of the AI community with it!

It is now becoming clear that the most important work making the transition from the AI labs to the standards of the World Wide Web is in the area of web ontologies. In the mid to late 1990s, several important projects showed the utility of tying machine-readable ontologies to resources on the web. These projects led to significant government interest in the area, and under the aegis of funding from the US DARPA and the EU's IST program, the Semantic Web began to grow – gaining in size, capability and interest by leaps and bounds. Mechanisms for embedding knowledge in the web are now being standardized, and industry is beginning to take significant notice of this emerging trend. As the CTO for software of a large multi-national corporation, Richard Hayes-Roth of Hewlett-Packard, put it 'we expect the Semantic Web to be as big a revolution as the original Web itself.' (*Business Week*, February 2002).

Comprised of many of the top European researchers working in the Ontology area, the On-To-Knowledge project, from where much of the work described in this book originates, is a major contributor to this coming revolution. The book sets out new approaches to the development and deployment of knowledge on the web, and sets a precedent for high quality research in this exciting new area. This collection thus portrays state-of-the-art work demonstrating the power of new approaches to online knowledge management.

In short, we now see the day when the careful encapsulation of knowledge into domain-specific applications is replaced by a ubiquity of knowledge sources linked together into a large, distributed web of knowledge. Databases, web services, and documents on the web will all be able to bring this power to bear – with machine-readable ontologies helping to power a new wave of applications. The projects described in this book are the harbingers of this coming revolution, the leading edge of this new version of the 'knowledge is power' revolution.

James Hendler
University of Maryland

Biographies

Dr John Davies

Head of Adanced Business Applications
British Telecommunications plc, UK
john.nj.davies@bt.com

John Davies graduated from the University of London with a degree in Physics. He obtained a Masters degree in Computer Science and a doctorate in Artificial Intelligence from the University of Essex. He joined BT in 1990 and currently leads the Advanced Business Applications team in BTexact Technologies, BT's R&D arm, where he has responsibility for work in the areas of eBusiness, mCommerce and Knowledge Management. He has been responsible for the development of a set of intranet-based knowledge management tools which have been successfully deployed within BT and are the subject of a number of patents. This has led to the setting up of a spin-off company, Exago, of which he is the CTO.

Dr Davies is a frequent speaker at conferences on knowledge management and he has authored and edited many papers and books in the areas of the Internet, intelligent information access and knowledge management. Current research interests include the Semantic Web, online communities of practice, intelligent WWW search and collaborative virtual environments.

He is a visiting lecturer at Warwick Business School. He is a Chartered Engineer and a member of the British Computer Society, where he sits on the Information Retrieval expert committee.

Professor Dieter Fensel

University of Innsbruck
Austria

Dieter Fensel obtained a Diploma in Social Science at the Free University of Berlin and a Diploma in Computer Science at the Technical University of Berlin in 1989. In 1993 he was awarded a Doctor's degree in economic science (Dr. rer. pol.) at the University of Karlsruhe and in 1998 he received his Habilitation in Applied Computer Science. He has worked at the University of Karlsruhe (AIFB), the University of Amsterdam (UvA), and the Vrije Universiteit Amsterdam (VU). Since 2002, he has been working at the University of Innsbruck, Austria. His current research interests include ontologies, semantic web, web services, knowledge management, enterprise application integration, and electronic commerce.

He has published around 150 papers as journal, book, conference, and workshop contributions. He has co-organized around 100 scientific workshops and conferences and has edited several special issues of scientific journals. He is Associate Editor of the Knowledge and Information Systems in 1989, IEEE Intelligent Systems, the Electronic Transactions on Artificial Intelligence (ETAI), and Web Intelligence and Agent Systems (WIAS). He is involved in many national and international research projects, and in particular has been the project coordinator of the EU Ontoknowledge, Ontoweb, and SWWS projects.

Dieter Fensel is the co-author of the books *Intelligent Information Integration in B2B Electronic Commerce*, Kluwer, 2002; *Ontologies: Silver Bullet for Knowledge Management and Electronic Commerce*, Springer-Verlag, Berlin, 2001; *Problem-Solving Methods: Understanding, Development, Description, and Reuse*, Lecture Notes on Artificial Intelligence (LNAI), no 1791, Springer-Verlag, Berlin, 2000; and *The Knowledge Acquisition and Representation Language KARL*, Kluwer Academic Publisher, Boston, 1995.

Professor Frank van Harmelen

Department of AI
Vrije Universtiteit Amsterdam
Netherlands

Frank van Harmelen (1960) is professor in Knowledge Representation and Reasoning at the Department of Artificial Intelligence of the Vrije Universiteit Amsterdam. He studied mathematics and computer science in Amsterdam. In 1989, he was awarded a PhD from the Department of AI in Edinburgh for his research on meta-level reasoning. After holding a post-doctorate position at the University of Amsterdam, he moved to the Vrije Universiteit Amsterdam, where he currently heads the Knowledge Representation and Reasoning research group. He is the author of a book on meta-level inference, and editor of a book on knowledge-based systems.

He has published over 60 papers, many of them in leading journals and conferences. He has made key contributions to the CommonKADS project by providing a sound formal basis for the conceptual models. More recently, he has been co-project manager of the OnToKnowledge project, and was one of the designers of OIL, which (in its form DAML+OIL) is currently the basis for a W3C standardized Web ontology language. He is a member of the joint EU/US committee on agent markup languages (who are designing DAML+OIL), and a member of the W3C working group on Web Ontology languages.

List of Contibutors

John Davies, Richard Weeks, Uwe Krohn, Alistair Duke and Audrius Stonkus
BTexact Technologies, Orion 5/12, Adastral Park, Ipswich IP5 3RE, UK
{john.nj.davies, richard.weeks, uwe.krohn, alistair.duke, audrius.stonkus}@bt.com
http://www.quizxml.com/people/johndavies/

Ian Horrocks
Department of Computer Science, University of Manchester, Kilburn Building, Oxford
Road, Manchester, M13 9PL, UK
ihorrocks@cs.man.ac.uk
http://www.cs.man.ac.uk/~horrocks/

Dieter Fensel
Universitaet Innsbruck
Technikerstrasse 25, A-6020 Innsbruck, Austria
dieter.fensel@uibk.ac.at
http://informatik.uibk.ac.at/users/c70385/

York Sure, Rudi Studer and Steffan Staab
Institute AIFB, University of Karlsruhe, 76128 Karlsruhe, Germany
{sure, studer, staab}@aifb.uni-karlsruhe.de
http://www.aifb.uni-karlsruhe.de/wbs/

Michael Erdmann
Ontoprise GmbH, Haid-und-Neu-Str. 7, 76131 Karlsruhe, Germany
erdmann@ontoprise.de
http://www.ontoprise.de/

Jeen Broekstra, Arjohn Kampman, Christiaan Fluit, Herko ter Horst, Jos van der Meer
AIdministrator nederland bv, Amersfoort, Netherlands
{jeen.broekstra, arjohn.kampman, christiaan.fluit, herko.ter.horst,
jmee}@aidministrator.nl
http://www.aidministrator.nl/

Atanas Kiryakov, Kiril Simov, Damyan Ognayov
OntoText Lab, Sirma AI Ltd.
38A Chr. Botev blvd, Sofia 1000, Bulgaria
{naso,kivs,damyan}@sirma.bg
http://www.sirma.bg/

Frank van Harmelen, Hans Akkermans, Ying Ding, Peter Mika, Michel Klein, Marta
Sabou, Boris Omelayenko
Division of Mathematics & Computer Science, Free University, Amsterdam,
De Boelelaan 1081a, 1081 HV Amsterdam, Netherlands
{frank.van.harmelen, hansakkermans, ying, pmika, michel.klein, marta,
boris}@cs.vu.nl
http://www.cs.vu.nl

Robert Engels, Till Christopher Lech
CognIT a.s., Meltzersgt. 4, 0254 Oslo, Norway
robert.engels@cognit.no
http://www.cognit.no

Ulrich Reimer, Peter Brockhausen, Thorsten Lau, Jacqueline Reich
Swiss Life, IT Research & Development, P.O. Box, CH-8022 Zürich, Switzerland
{ulrich.reimer, peter.brockhausen, thorsten.lau, jacqueline.reich}@swisslife.ch
http://www.swisslife.ch

Victor Iosif
EnerSearch AB, Malmo, Sweden
victor@enersearch.se
http://www.enersearch.se/

Rikard Larsson
Lund University Business School,
Lund University, Box 7080, 22007 Lund
Sweden
rikard.larsson@fek.lu.se
http://www.lu.se/lu/engindex.html

Acknowledgements

Cath McCarney is thanked for her significant contribution to the typographical preparation of this volume.

Chapter 3: Hans-Peter Schnurr, Hans Akkermans and colleagues from AIFB, University of Karlsruhe, are thanked.

Chapter 8: The authors would like to mention Dirk Wenke, Siggi Handschuh and Alexander Mädche , who implemented large parts of the OntoEdit Ontology Engineering Environment, and Jürgen Angele and Steffen Staab, who contributed valuable input for this work.

Chapter 10: Nick Kings is thanked for his contribution to the design and development of the OntoShare system.

Chapter 13: The authors thank their former colleagues Bernd Novotny and Martin Staudt who put considerable effort into earlier phases of the two case studies described in this chapter.

The work in this book has been partially supported by the European Commission research project OnToKnowledge (IST-1999-10132), and by the Swiss Federal Office for Education and Science (project number BBW 99.0174). Vincent Obozinski, Wolfram Brandes, Robert Meersman and Nicola Guarino are thanked for their constructive feedback on the On-To-Knowledge project.

Elisabeth, Joshua and Thomas – thanks for the patience and the inspiration.
JD.

1

Introduction

John Davies, Dieter Fensel and Frank van Harmelen

There are now several billion documents on the World Wide Web (WWW), which are used by more than 300 million users globally, and millions more pages on corporate intranets. The continued rapid growth in information volume makes it increasingly difficult to find, organize, access and maintain the information required by users. The notion of a Semantic Web (Berners-Lee et al., 2001) that provides enhanced information access based on the exploitation of machine-processable meta-data has been proposed. In this book, we are particularly interested in the new possibilities afforded by Semantic Web technology in the area of knowledge management.

Until comparatively recently, the value of a company was determined mainly by the value of its tangible assets. In recent years, however, it has been increasingly recognized that in the post-industrial era, an organization's success is more dependent on its intellectual assets than on the value of its physical resources.

This increasing importance of intangible assets is evident from the high premiums on today's stockmarkets. We can measure this by expressing the market value of a company as a percentage of its book value. Looking at this index, we see that the Dow Jones Industrial has risen steadily over the last 25 years and now stands at around 300%, notwithstanding recent stockmarket falls.

Underlying this trend are a number of factors. The requirement for highly-skilled labour in many industries, new computing and telecommunications technologies, faster innovation and ever shorter product cycles, has caused a huge change in the ways organizations compete: *knowledge is now the key battleground for competition.*

Other factors driving companies to try and manage and exploit their intellectual assets more effectively are: increasing employee turnover rates and a more mobile workforce, which can lead to loss of knowledge; and globalization, often requiring people to collaborate and exchange knowledge across continents and time zones.

The knowledge management discipline aims to address this challenge and can be broadly defined as the tools, techniques and processes for the most effective and efficient management of an organization's intellectual assets (Davies, 2000a). These intellectual assets can be exploited in a variety of ways. By sharing and re-using current best practice, for instance, current business processes can be improved, and duplication of effort can be eliminated. New business opportunities can be generated by collecting intelligence on markets and sales leads; and new products and services can be created, developed and brought to the marketplace ahead of competitors.

It is often argued in knowledge management circles that technology is a relatively marginal aspect of any knowledge management initiative and that organizational culture is far more important. While the sentiment that we need a wider perspective than just technology is correct, this viewpoint reveals the assumption of a dichotomy between technology and organizational culture which does not exist. Rather, technology-based tools are among the many artefacts entwined with culture, whose use both affects and is affected by the prevailing cultural environment. A holistic view is required and technology often plays a larger part in cultural factors than is sometimes acknowledged. Although the focus of this book is Semantic Web-based tools for knowledge management, it is equally important to understand the cultural and organizational contexts in which such tools can be used to best effect. Related work in this area can be found, for example, in Maxwell (2000).

1.1 The Semantic Web and Knowledge Management

Intranets have an important role to play in the more effective exploitation of both explicit (codified) and tacit (unarticulated) knowledge. With regard to explicit knowledge, intranet technology provides a ubiquitous interface to an organization's knowledge at relatively low cost using open standards. Moving information from paper to the intranet can also have benefits in terms of speed of update and hence accuracy. The issue then becomes how to get the right information to the right people at the right time: indeed, one way of thinking about explicit knowledge is that it is information in the right context; that is, information which can lead to effective action. With tacit knowledge, we can use intranet-based tools to connect people with similar interests or concerns, thus encouraging dialogue and opening up the possibility of the exchange of tacit knowledge.

Important information is often scattered across web and/or intranet resources. Traditional search engines return ranked retrieval lists that offer little or no information on the semantic relationships among documents. Knowledge workers spend a substantial amount of their time browsing and reading to find out how documents are related to one another and where each falls into the overall structure of the problem domain. Yet only when knowledge workers begin to locate the similarities and differences among pieces of information do they move into an essential part of their work: building relationships to create new knowledge.

Current knowledge management systems have significant weaknesses:

- *Searching information*: existing keyword-based searches can retrieve irrelevant information that includes certain terms in different meanings. They also miss information when different terms with the same meaning about the desired content are used. Information retrieval traditionally focuses on the relationship between a given query (or user profile) and the information store. On the other hand, exploitation of interrelationships between selected pieces of information (which can be facilitated by the use of ontologies) can put otherwise isolated information into a meaningful context. The implicit structures so revealed help users use and manage information more efficiently (Davies, 1999).
- *Extracting information*: currently, human browsing and reading is required to extract relevant information from information sources. This is because automatic agents do not possess the common sense knowledge required to extract such information from textual representations, and they fail to integrate information distributed over different sources.
- *Maintaining* weakly structured text sources is a difficult and time-consuming activity when such sources become large. Keeping such collections consistent, correct, and up-to-date requires mechanized representations of semantics that help to detect anomalies.
- *Automatic document generation* would enable adaptive websites that are dynamically reconfigured according to user profiles or other aspects of relevance. Generation of semi-structured information presentations from semi-structured data requires a machine-accessible representation of the semantics of these information sources.

The competitiveness of many companies depends heavily on how they exploit their corporate knowledge and memory. Most networked information is now typically multimedia and rather weakly structured. This is not only true of the Internet but also of large company intranets. Finding and maintaining information is a challenging problem in weakly structured representation media. Increasingly, companies have realized that their intranets are valuable repositories of corporate knowledge. But as volumes of information continue to increase rapidly, the task of turning this resource into useful knowledge has become a major problem.

Knowledge management tools are needed that integrate the resources dispersed across web resources into a coherent corpus of interrelated information. Previous research in information integration (see, e.g., Hearst, 1998) has largely focused on integrating heterogeneous databases and knowledge bases, which represent information in a highly structured way, often by means of formal languages. In contrast, the web consists to a large extent of unstructured or semi-structured natural language text.

The *Semantic Web* is envisioned as an extension of the current web where, in addition to being human-readable using WWW browsers, documents are annotated with meta-information. This meta-information defines what the information (documents) is about in a machine processable way. The explicit representation of meta-information, accompanied by domain theories (i.e. ontologies), will enable a web that provides a qualitatively new level of service. It will weave together an incredibly large network of human knowledge and will complement it with machine processability. Various automated services will help the user achieve goals by accessing and providing information in machine-understandable form. This process may ultimately create extremely knowledgeable systems with various specialized reasoning services systems that can support us in nearly all aspects of life and that will become as necessary to us as access to electric power.

Ontologies offer a way to cope with heterogeneous representations of web resources. The domain model implicit in an ontology can be taken as a unifying structure for giving information a common representation and semantics.

1.2 The Role of Ontologies

Ontologies are a key enabling technology for the Semantic Web. They interweave human understanding of symbols with their machine processability. Ontologies were developed in artificial intelligence to facilitate knowledge sharing and re-use. Since the early 1990s, ontologies have become a popular research topic. They have been studied by several artificial intelligence research communities, including knowledge engineering, natural-language processing and knowledge representation. More recently, the use of ontologies has also become widespread in fields such as intelligent information integration, cooperative information systems, information retrieval, electronic commerce, and knowledge management. The reason ontologies are becoming popular is largely due to what they promise: *a shared and common under-*

standing of a domain that can be communicated between people and applica-tion systems. As such, the use of ontologies and supporting tools offers an opportunity to significantly improve knowledge management capabilities in large organizations and it is their use in this particular area which is the subject of this book.

It describes a Semantic Web-based knowledge management architecture and a suite of innovative tools for semantic information processing. The theoretical underpinnings of our approach are also set out. The tool environ-ment addresses three key aspects:

- Acquiring ontologies and linking them with large amounts of data. For reasons of scalability this process must be automated based on information extraction and natural language processing technology. For reasons of quality this process requires the human in the loop to build and manipulate ontologies using ontology editors.
- Storing and maintaining ontologies and their instances. We developed a resource description framework (RDF) schema repository that provides database technology and simple forms of reasoning over web information sources.
- Querying and browsing semantically enriched information sources. We describe semantically enriched search engines, browsing and knowledge sharing support that makes use of machine processable semantics of data.

The developed technology has been proven to be useful in a number of case studies. We discuss improved information access in the intranet of a large organization (Lau and Sure, 2002). The technology has also been used to facilitate electronic knowledge sharing and reuse in a technology firm and knowledge management in a virtual organization. We now move to a more detailed discussion of our architecture.

1.3 An Architecture for Semantic Web-based Knowledge Management

Figure 1.1 shows our architecture for knowledge management based on the Semantic Web. The architecture addresses all the key stages of the knowledge management lifecycle (with one exception – the methodology, which we mention shortly):

1.3.1 Knowledge Acquisition

Given the large amounts of unstructured and semi-structured information held on organizational intranets, automatic knowledge *extraction from unstruc-*

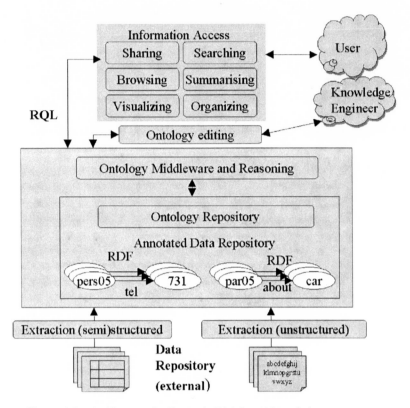

Figure 1.1   Architecture for Semantic Web-based knowledge management

tured and semi-structured data in external data repositories is required and this is shown in the bottom layer of the diagram. Support for human knowledge acquisition is also needed and the knowledge engineer needs to be supported by *ontology editing tools* which support the creation, maintenance and population of ontologies.

1.3.2 Knowledge Representation

Once knowledge has been acquired from human sources or automatically extracted, it is then required to represent the knowledge in an ontology language (and of course to provide a query language to provide access to the knowledge so stored). This is the function of the *ontology repository*.

1.3.3 Knowledge Maintenance

Ontology middleware is required with support for development, management, maintenance, and use of knowledge bases.

1.3.4 Knowledge Use

Finally, and perhaps most importantly, *information access* tools are required to allow end users to exploit the knowledge represented in the system. Such tools include facilities for finding, sharing, summarizing, visualizing, browsing and organizing knowledge.

1.4 Tools for Semantic Web-based Knowledge Management

Figure 1.2 makes this diagram more concrete by instantiating the various modules of the abstract architecture with a number of tools which are

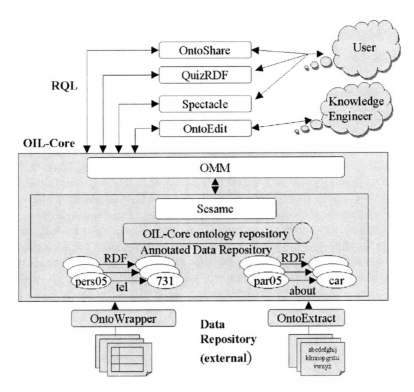

Figure 1.2   Tools for Semantic Web-based knowledge management

described in later chapters. Here we briefly mention each tool and the chapter in which it is described.

1.4.1 Knowledge Acquisition

OntoWrapper for knowledge extraction from semi-structured information and OntoExtract, which extracts meta-data from unstructured information are discussed in Chapter 6. Support for human knowledge acquisition is discussed in the context of the OntoEdit system in Chapter 7, which supports the creation, maintenance and population of ontologies in a variety of data formats.

1.4.2 Knowledge Representation

A fully-fledged RDF data repository (the SESAME system) is described in Chapter 5. In addition to data storage, SESAME supports RDF querying in two leading RDF query languages.

1.4.3 Knowledge Maintenance

The ontology middleware module (OMM) is described in Chapter 11.

1.4.4 Knowledge Use

A range of information access tools for the Semantic Web are discussed. QuizRDF, described in Chapter 8, is a semantic search engine for browsing and querying RDF-annotated information resources. Spectacle (Chapter 9) is a visualization and browsing tool for ontology-based information. Chapter 10 describes OntoShare, an RDF-based system which supports knowledge sharing between users, using Semantic Web technology to create an ontology-based information resource automatically from the information so shared.

In addition to the chapters outlined above, the book contains seven further chapters in addition to this introduction. Chapter 2 discusses the pyramid of languages that underpin the Semantic Web. XML, RDF and RDF Schema are covered briefly and the chapter then focuses on OIL and DAML+OIL, currently the most prominent ontology languages for the Semantic Web.

Key to applying Semantic Web technology in the knowledge management arena is the development of appropriate ontologies for the domain and application at hand. Chapter 3 presents a five step methodology for application-driven ontology development. Once created, ontologies must of course be managed: they need to be stored, aligned, maintained and their evolution tracked. This important topic is the subject of Chapter 4.

Chapters 12 and 13 look at specific case studies using the tools and techniques described in earlier chapters. Chapter 12 covers two case studies from the Swiss Life insurance group in the application areas of skills management and intelligent information access in the domain of international accounting standards. Chapter 13 looks at the application of Semantic Web tools for knowledge dissemination in a virtual organization.

Chapter 14 looks ahead to the future potential of the emergence and combination of the P2P computing paradigm and the use of Semantic Web technologies. In Chapter 15, we offer some brief concluding remarks and consider prospects for a truly global Semantic Web.

2

OIL and DAML+OIL: Ontology Languages for the Semantic Web

Dieter Fensel, Frank van Harmelen and Ian Horrocks

2.1 Introduction

This chapter discusses OIL and DAML+OIL, currently the most prominent ontology languages for the Semantic Web.

The chapter starts by discussing the pyramid of languages that underlie the architecture of the Semantic Web (XML, RDF, RDFS). In section 2.2, we briefly describe XML, RDF and RDFS. We then discuss in more detail OIL and DAML+OIL, the first proposals for languages at the ontology layer of the semantic pyramid. For OIL (and to some extent DAML+OIL) we discuss the general design motivations (Section 2.3), describe the constructions in the language (Section 2.4), and the various syntactic forms of these languages (Section 2.5). Section 2.6 discusses the layered architecture of the language, section 2.7 briefly mentions the formal semantics, section 2.8 discusses the transition from OIL to DAML+OIL, and section 2.9 concludes with our experience with the language to date and future development in the context of the World Wide Web Consortium (W3C).

This chapter is not intended to give full and formal definitions of either the syntax or the semantics of OIL or DAML+OIL. Such definitions are already available elsewhere: http://www.ontoknowledge.org/oil/ for OIL and http://www.w3.org/submission/2001/12/ for DAML+OIL.

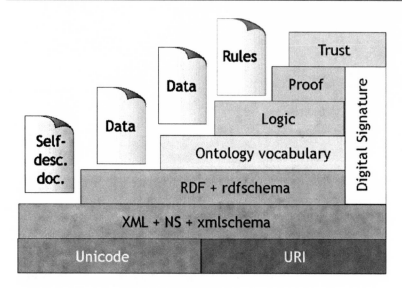

Figure 2.1

2.2 The Semantic Web Pyramid of Languages

One of the main architectural premises of the Semantic Web is a stack of languages, often drawn in a figure first presented by Tim Berners-Lee in his XML 2000 address (http://www.w3.org/2000/talks/1206-xml2k-tbl/slide1-0. html) (see Figure 2.1). We briefly discuss all of the layers in this language stack leading up to the ontology languages.

2.2.1 XML for Data Exchange

XML is already widely known, and is the basis for a rapidly growing number of software development activities. It is designed for mark-up in documents of arbitrary structure, as opposed to HTML, which was designed for hypertext documents with fixed structures. A well-formed XML document creates a balanced tree of nested sets of open and close tags, each of which can include several attribute-value pairs. There is no fixed tag vocabulary or set of allowable combinations, so these can be defined for each application. In XML 1.0 this is done using a document type definition (DTD) to enforce constraints on which tags to use and how they should be nested within a document. A DTD defines a grammar to specify allowable combinations and nesting of tag names, attribute names, and so on. Developments are well underway at W3C to replace DTDs with XML Schema definitions. Although XML Schema offers several advantages over DTDs, their role is essentially the same: to define a grammar for XML documents.

XML is used to serve a range of purposes:

- Serialization syntax for other mark-up languages. For example, the synchronized multimedia integration language (SMIL) is syntactically just a particular XML DTD; it defines the structure of a SMIL document. The DTD is useful because it facilitates a common understanding of the meaning of the DTD elements and the structure of the DTD.
- Separating form from content. An XML serialization can be used in a web page with an XSL style sheet to render the different elements appropriately.
- Uniform data-exchange format. An XML serialization can also be transferred as a data object between two applications.

It is important to note that in all these applications of XML, a DTD (or an XML schema) only specifies syntactic conventions; any intended semantics are outside the realm of the XML specification.

2.2.2 RDF for Assertions

The resource description framework (RDF) is a recent W3C recommendation designed to standardize the definition and use of meta-data descriptions of web-based resources. However, RDF is equally well suited to representing data.

The basic building block in RDF is an object–attribute–value triple, commonly written as A(O,V). That is, an object O has an attribute A with value V. Another way to think of this relationship is as a labelled edge between two nodes:

```
[O]-A → [V]
```

This notation is useful because RDF allows objects and values to be interchanged. Thus, any object can play the role of a value, which amounts to chaining two labelled edges in a graphic representation. Figure 2.2, for example, expresses the following three relationships in A(O,V) format:

```
hasName('http://www.w3.org/employee/id1321',
        'Jim Lerners')
authorOf('http://www.w3.org/employee/id1321',
         'http://www.books.org/ISBN0062515861')
hasPrice('http://www.books.org/ISBN0062515861',
         "$62").
```

RDF uses XML as its serialization syntax (i.e. using XML in the first of its

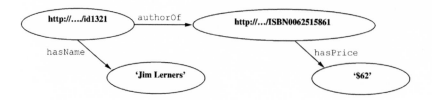

Figure 2.2

intended uses listed above). The first of the three A(O,V) triples would look as follows in RDF's XML serialization:

```
<rdf:Description rdf:about=
    "http://www.w3.org/employee/id1321">
  <hasName rdf:resource="Jim Lerners"/>
</rdf:Description>
```

RDF also allows a form of reification in which any RDF statement can be the object or value of a triple, which means graphs can be nested as well as chained. On the web this allows us, for example, to express doubt or support of statements created by other people. Finally, it is possible to indicate that a given object is of a certain type, such as stating that 'ISBN0012515866' is of the rdf:type book, by creating a type arc referring to the book definition in RDFS:

```
<rdf:Description rdf:about=
    "http://www.books.org/ISBN0062515861">
  <rdf:type rdf:resource=
    "http://description.org/schema/book">
</rdf:Description>
```

It is important to note that RDF is designed to provide a basic object–attribute–value data model for meta-data. Other than this intended semantics, described only informally in the standard, RDF makes no data-modelling commitments. In particular, no reserved terms are defined for further data modelling. As with XML, the RDF data model provides no mechanisms for declaring property names that are to be used.

2.2.3 RDF Schema for Simple Ontologies

RDF Schema takes a step further into a richer representation formalism and introduces basic ontological modelling primitives into the web. With RDFS, we can talk about classes, subclasses, subproperties, domain and range restrictions of properties, and so forth in a web-based context.

Despite the similarity in their names, RDFS fulfils a different role than XML Schema. XML Schema, and also DTDs, prescribe the order and combination of tags in an XML document. In contrast, RDFS only provides information about the interpretation of the statements given in an RDF data model, but it does not constrain the syntactical appearance of an RDF description.

RDFS lets developers define a particular vocabulary for RDF data (such as hasName) and specify the kinds of object to which these attributes can be applied. In other words, the RDFS mechanism provides a basic type system for RDF models. This type system uses some predefined terms, such as `Class`, `subPropertyOf`, and `subClassOf`. RDFS expressions are also valid RDF expressions (just as XML Schema expressions are valid XML). RDF objects can be defined as instances of one or more classes using the `type` property. The `subClassOf` property allows the developer to specify the hierarchical organization of such classes:

```
<rdfs:Class rdf:about="Book"/>

<rdfs:Class rdf:about="HardCover">
 <rdfs:subClassOf rdf:resource="#Book"/>
</rdfs:Class>

<Hardcover rdf:resource=
 "http://www.books.org/ISBN0062515861"/>
```

Properties can be defined with their `domain` and `range`, and they can be organized in a property hierarchy using `subPropertyOf`:

```
<rdfs:Property rdf:about="hasPrice">
 <rdfs:domain rdf:resource="#Book"/>
</rdfs:Property>
```

2.3 Design Rationale for OIL

The previous section shows that RDFS can be regarded as a very simple ontology language. However, many types of knowledge cannot be expressed in this simple language. Just a few examples of useful things we cannot say in RDFS are:

- stating that every book has exactly one price, but at least one author (and possibly more);
- stating that titles of books are strings and prices of books are numbers;

- stating that no book can be both hardcover and softcover;
- stating that every book is either hardcover or softcover (i.e. there is no other option than these two).

It is clear that a richer language than RDFS is required if we want to be able to express anything but the most trivial domain models on the Semantic Web. OIL aims to be such a language.

The following have been important design goals for OIL:

- maximizing compatibility with existing W3C standards, such as XML and RDF;
- maximizing partial interpretability by less semantically aware processors;
- providing modelling primitives that have proven useful for large user communities;
- maximizing expressiveness to enable modelling of a wide variety of ontologies;
- providing a formal semantics (a mathematically precise description of the meaning of every expression) in order to facilitate machine interpretation of that semantics;
- enabling sound, complete and efficient reasoning services, if necessary by limiting the expressiveness of the language.

These design goals lead to the following three requirements:

- It must be highly intuitive to the human user. Given the success of the frame-based and object-oriented modelling paradigm, an ontology should have a frame-like look and feel.
- It must have a well-defined formal semantics with established reasoning properties to ensure completeness, correctness, and efficiency.
- It must have a proper link with existing web languages such as XML and RDF to ensure interoperability.

We now discuss each of these three requirements briefly.

2.3.1 Frame-based Systems

The central modelling primitives of predicate logic are relations (predicates). Frame-based and object-oriented approaches take a different viewpoint. Their central modelling primitives are classes (or frames) with certain properties called attributes. These attributes do not have a global scope but apply only to the classes for which they are defined; we can associate the same attribute name with different range restrictions when defined for different classes. A frame provides a context for modelling one aspect of a domain. Researchers have developed many other additional refinements of these modelling constructs, which have led to this modelling paradigm's success. Many frame-based systems and languages have emerged, and, renamed as object orientation, they have conquered the software engineering community. OIL incorporates the essential modelling primitives of frame-based systems: it is based on the notion of a concept and the definition of its superclasses and attributes. Relations can also be defined not as an attribute of a class but as an independent entity having a certain domain and range. Like classes, relations can fall into a hierarchy. OIL's modelling primitives are further discussed in Section 2.4.

2.3.2 Description Logics

Description logics (DL) describes knowledge in terms of concepts and role restrictions that can automatically derive classification taxonomies. The main thrust of knowledge representation research is to provide theories and systems for expressing structured knowledge and for accessing and reasoning with it in a principled way. In spite of the discouraging theoretical worst-case complexity of the results, there are now efficient implementations for DL languages, which we explain later. OIL inherits from DL its formal semantics and the efficient reasoning support. The semantics of OIL are briefly discussed in Section 2.7.

2.3.3 Web Standards: XML and RDF

Modelling primitives and their semantics are one aspect of an ontology language, but we still have to decide about its syntax. Given the web's current dominance and importance, we must formulate a syntax of an ontology exchange language with existing web standards for information representation. First, OIL has a well-defined syntax in XML based on a DTD and an XML Schema definition. Second, OIL is an extension of RDF and RDFS. With regard to ontologies, RDFS provides two important contributions: a standardized syntax for writing ontologies and a standard set of modelling primitives such as instance-of and subclass-of relationships. OIL's relation to XML and RDF(S) is discussed in Section 2.5.

2.4 OIL Language Constructs

The frame structure of OIL is based on XOL (Karp et al., 1999), an XML serialization of the OKBC-lite knowledge model (Chaudhri et al., 1998). In these languages classes (concepts) are described by frames, whose main components consist of a list of superclasses and a list of slot-filler pairs.

OIL extends this basic frame syntax so that it can capture the full power of an expressive description logic. These extensions include the following:

- Arbitrary Boolean combinations of classes (called class expressions) can be formed, and used anywhere that a class name can be used. In particular, class expressions can be used as slot fillers, whereas in typical frame languages slot fillers are restricted to being class (or individual) names.
- A slot-filler pair (called a slot constraint) can itself be treated as a class: it can be used anywhere that a class name can be used, and can be combined with other classes in class expressions.
- Class definitions (frames) have an (optional) additional field that specifies whether the class definition is primitive (a subsumption axiom) or non-primitive (an equivalence axiom). If omitted, this defaults to primitive.
- Different types of slot constraint are provided, specifying value restriction, existential quantification and various kinds of cardinality constraint (some frame languages also provide this feature, referring to such slot constraints as facets).
- Global slot definitions are extended to allow the specification of superslots (subsuming slots) and of properties such as transitive and symmetrical.
- Unlike many frame languages, there is no restriction on the ordering of class and slot definitions, so classes and slots can be used before they are 'defined''. This means that OIL ontologies can contain cycles.
- In addition to standard class definitions (frames), OIL also provides axioms for asserting disjointness, equivalence and coverings with respect to class expressions (and not just with respect to atomic concepts).

Many of these points are standard for a DL, but are novel for a frame language.

OIL is also more restrictive than typical frame languages in some respects. In particular, it does not support collection types other than sets (e.g. lists or bags), and it does not support the specification of default fillers. These restrictions are necessary in order to maintain the formal properties of the language (e.g. monotonicity) and the correspondence with description logics.

2.4.1 A Simple Example in OIL

Below is a simple example of an OIL ontology taken from a case-study at Swiss Life for constructing an ontology-based skills-management system.

```
begin-ontology

ontology-container

title "Swiss Life skills DB"
```

```
creator "Ullrich Reimer"
description "Part of the ontology from the Swiss Life
            case study"
description.release "1.0"
type ontology
identifier "SLSkillsDB"
language "OIL"
```

ontology-definitions

```
class-def Department
instance-of ITDept Department
class-def Skills
  slot-constraint SkillsLevel cardinality 1
slot-def HasSkills
  domain Employee
  range Skills
slot-def WorksInProject
  domain Employee
  range Project
  inverse ProjectMembers
class-def defined ITProject
  subclass-of Project
  slot-constraint ResponsibleDept has-value ITDept
slot-def ManagementLevel
  domain Employee
  range one-of "member" "head-of-group"
  "head-of-dept" "CEO"
class-def Publishing
  subclass-of Skills
class-def DocumentProcessing
  subclass-of Skills
class-def DesktopPublishing
  subclass-of Publishing and DocumentProcessing
instance-of GeorgeMiller Employee
related HasSkills GeorgeMiller
        DesktopPublishingLevel3
instance-of DesktopPublishingLevel3
            DesktopPublishing
related SkillsLevel DesktopPublishingLevel3 3
```

end-ontology

This is a snapshot from a larger ontology defined at Swiss Life for a skills-management case study. Every OIL ontology is itself annotated with meta-data, starting with such things as title, creator, creation date, and so on. OIL follows the W3C Dublin Core Standard on bibliographical meta-data for this purpose. Any ontology language's core is its hierarchy of class declarations, stating, for example, that Department is a class, and that ITDept is an instance of that class. Skills are another class, this time with an associated slot SkillsLevel. The cardinality constraint stipulates that every Skill must have exactly one SkillsLevel. Skills are the range of a relation HasSkills (between Employees and Skills). WorksInProject is another relation defined on Employees (i.e. another slot of the Employee class. ProjectMembers is defined as the inverse relation of WorksInProject. Projects come in various subclasses, one of which is ITProject. ITProjects are exactly those Projects whose ResponsibleDept slot has at least the value ITDept. A third slot defined on Employees is their ManagementLevel. Values for this slot are restricted to one of the enumerated values. Next, two subclasses of Skills are defined (Publishing and DocumentProcessing). The class DesktopPublishing is defined to be exactly the intersection of both of these two skills. Finally GeorgeMiller is defined to be particular Employees who has a DesktopPublishing skill of SkillsLevel 3.

2.5 Different Syntactic Forms

The above language description uses OIL's 'human readable' serialization. This aids readability, but is not suitable for publishing ontologies on the web. For this purpose OIL is also provided with both XML and RDFS serializations. OIL's XML serialization directly corresponds with the human readable form. Its main benefit is to provide a format that is easier to parse than the more human-readable form shown above. A full specification in the form of an XML DTD and XML Schema can found on the OIL website (http://www.on-toknowledge.org/oil).

The RDFS serialization is more interesting as it uses the features of RDFS to capture as much as possible of OIL ontologies in RDFS. The following code shows part of the RDFS serialization of the skills-management example given above:

```
<rdf:Description rdf:about=""
 dc:creator="Ullrich Reimer"
 dc:description.release="1.0"
 dc:language="OIL"
 dc:title="Swiss Life skills DB"
```

```
dc:type="ontology">
<dc:description>"Part of the ontology from the
Swiss Life case study"</dc:description>
</rdf:Description>
<rdfs:Class rdf:about="DesktopPublishing">
 <rdfs:subClassOf>
   <oil:And>
     <oil:hasOperand rdf:resource=
     "DocumentProcessing"/>
     <oil:hasOperand rdf:resource="Publishing"/>
   </oil:And>
 </rdfs:subClassOf>
</rdfs:Class>
<DesktopPublishing rdf:about=
"DesktopPublishingLevel3">
 <SkillsLevel rdf:resource="3"/>
</DesktopPublishing>
<rdfs:Class rdf:about="DocumentProcessing">
 <rdfs:subClassOf rdf:resource="Skills"/>
</rdfs:Class>
<Employee rdf:about="GeorgeMiller">
 <HasSkills rdf:resource=
 "DesktopPublishingLevel3"/>
</Employee>
<rdf:Property rdf:about="HasSkills">
 <rdfs:domain rdf:resource="Employee"/>
 <rdfs:range rdf:resource="Skills"/>
</rdf:Property>
<oil:DefinedClass rdf:about="ITProject">
 <rdfs:subClassOf rdf:resource="Project"/>
 <oil:hasPropertyRestriction>
   <oil:HasValue>
     <oil:toClass rdf:resource = "ITDept"/>
     <oil:onProperty rdf:resource=
     "ResponsibleDept"/>
   </oil:HasValue>
 </oil:hasPropertyRestriction>
</oil:DefinedClass>

</rdf:RDF>
```

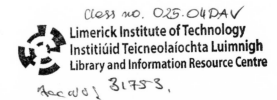

Class no. 025.04 DAV
Limerick Institute of Technology
Institiúid Teicneolaíochta Luimnigh
Library and Information Resource Centre

Acc No 31753.

The RDFS serialization makes clear that OIL's ontology-container is indeed expressed using Dublin Core properties. It also shows that OIL's RDFS form re-uses as much as possible the constructions already available in RDFS, such as `rdfs:Class`, `rdfs:domain`, `rdfs:range`, `rdf:Property`, etc. The main value of this is to make OIL ontologies accessible to software that only understands the weaker RDFS language. More on the usefulness of this in the next section.

The RDFS serialization also attempts to define a 'meta-ontology' describing the structure of the OIL language itself. The RDFS code below shows part of the RDFS description of OIL.

```
<rdfs:Class rdf:ID="DefinedClass">
  <rdfs:subClassOf rdf:resource=
    "http://www.w3.org/2000/01/rdf-schema#Class"/>
</rdfs:Class>
<rdf:Property rdf:ID="hasPropertyRestriction">
  <rdf:type rdf:resource=
    "http://www.w3.org/2000/01/
    rdf-schema#ConstraintProperty"/>
  <rdfs:domain rdf:resource=
    "http://www.w3.org/2000/01/rdf-schema#Class"/>
  <rdfs:range rdf:resource=
    "#PropertyRestriction"/>
</rdf:Property>

<rdfs:Class rdf:ID="PropertyRestriction">
  <rdfs:subClassOf rdf:resource=
    "#ClassExpression"/>
</rdfs:Class>

<rdfs:Class rdf:ID="HasValue">
  <rdfs:subClassOf rdf:resource=
    "#PropertyRestriction"/>
</rdfs:Class>

<rdf:Property rdf:ID="onProperty">
  <rdfs:domain rdf:resource=
    "#PropertyRestriction"/>
  <rdfs:range rdf:resource=
    "http://www.w3.org/1999/02/22-rdf-syntax-
    ns#Property"/>
</rdf:Property>
```

```
<rdf:Property rdf:ID="toClass'´>
  <rdfs:domain rdf:resource=
    "#PropertyRestriction"/>
  <rdfs:range rdf:resource=
    *"#ClassExpression"/>
</rdf:Property>

</rdf:RDF>
```

The 'meta-ontology' defines definedClass as a subclass of rdfs:Class. It also defines hasPropertyRestriction as an instance of RDFS ConstraintProperty that connects an RDFS class (the property's domain) to an OIL property restriction (the property's range). (Property is the RDF name for a binary relation like a slot or role). A PropertyRestriction (slot constraint) is then defined as a kind of ClassExpression, with HasValue (an existential quantification) being a kind of PropertyRestriction. Properties onProperty and toClass are then defined as 'meta-slots' of PropertyRestriction whose fillers will be the name of the property (slot) to be restricted and the restriction class expression. Again, all this helps to make OIL ontologies partly available to RDFS-only agents.

The complete description of OIL in RDFS, as well as a more detailed description of RDF and RDFS, can be found in Horrocks et al. (2000).

2.6 Language Layering

It is unlikely that a single ontology language can fulfil all the needs of the large range of users and applications of the Semantic Web. We have therefore organized OIL as a series of ever increasing layers of sublanguages. Each additional layer adds functionality and complexity to the previous layer. This is done such that agents (humans or machines) who can only process a lower layer can still partially understand ontologies that are expressed in any of the higher layers. A first and very important application of this principle is the relation between OIL and RDFS (Figure 2.3).

- **Core OIL** coincides largely with RDFS (with the exception of the reification features of RDFS). This means that even simple RDFS agents are able to process the OIL ontologies, and pick up as much of their meaning as possible with their limited capabilities.
- **Standard OIL** is a language intended to capture the necessary mainstream modelling primitives that both provide adequate expressive power and are

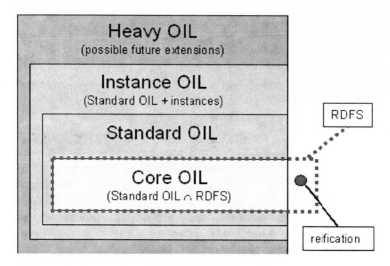

Figure 2.3

well understood, thereby allowing the semantics to be precisely specified
and complete inference to be viable.

- **Instance OIL** includes a thorough individual integration. While the
  previous layer – Standard OIL – included modelling constructs that
  allow individual fillers to be specified in term definitions, Instance OIL
  includes a full-fledged database capability.
- **Heavy OIL** may include additional representational (and reasoning)
  capabilities. Especially a more expressive rule languages and meta-class
  facilities seem highly desirable.

The layered architecture of OIL has three main advantages:

- First, an application is not forced to work with a language that offers
  significantly more expressiveness and complexity than is actually needed.
- Second, applications that can only process a lower level of complexity are
  still able to catch same of the aspects of an ontology.
- Third, an application that is aware of a higher level of complexity can still
  also understand ontologies expressed in a simpler ontology language.

Defining an ontology language as an extension of RDFS means that every
RDFS ontology is a valid ontology in the new language (i.e. an OIL processor will
also understand RDFS). However, the other direction is also available: defining
an OIL extension as close as possible to RDFS allows maximal reuse of existing
RDFS-based applications and tools. However, since the ontology language

usually contains new aspects (and therefore new vocabulary, which an RDFS processor does not know), 100% compatibility is not possible. If we look at the RDFS serialization of the Swiss Life skills-management ontology above, all statements that do not include the `oil:` namespace are already valid RDFS. It is noteworthy that this is already the majority of the ontology! Furthermore, even the statements that do include OIL-specific constructs are at least partially understandable to an RDFS processor. For example, the statement:

```
<rdfs:Class rdf:about="DesktopPublishing">
  <rdfs:subClassOf>
    <oil:And>
      <oil:hasOperand rdf:resource=
      "DocumentProcessing"/>
      <oil:hasOperand rdf:resource="Publishing"/>
    </oil:And>
  </rdfs:subClassOf>
</rdfs:Class>
```

still defines `DesktopPublishing` as a class for any RDFS processor, even though the relationship between `DesktopPublishing` and the classes `DocumentProcessing` and `Publishing` will be lost.

The value of the meta-ontology (which defines parts of OIL's semantics in terms of RDFS itself) becomes clear if we look at the statement:

```
<oil:DefinedClass rdf:about="ITProject">
  <rdfs:subClassOf rdf:resource="Project"/>
  <oil:hasPropertyRestriction>
    <oil:HasValue>
      <oil:toClass rdf:resource="ITDept"/>
      <oil:onProperty rdf:resource=
      "ResponsibleDept"/>
    </oil:HasValue>
  </oil:hasPropertyRestriction>
</oil:DefinedClass>
```

The meta-ontology states that `oil:DefinedClass` is a subclass of `rdfs:Class`. Since `ITProject` is defined to be an instance of `oil:DefinedClass`, it must also be an instance of `rdfs:Class`, and hence an RDFS processor would be able to recognize that `ITProject` is a class, even though this statement is almost fully couched in OIL specific vocabulary.

2.7 Semantics

The semantics of OIL relies on a translation into the SHIQ(D) description logic. SHIQ(D) has a highly expressive concept language that is able to fully capture the OIL core language, and a satisfiability preserving translation can be defined that maps OIL ontologies into SHIQ(D) terminologies. This has the added benefit that an existing SHIQ reasoner implemented in the FaCT system can be used to reason with OIL ontologies. The full details of this mapping have been given in Horrocks et al. (2000), and are not repeated here.

2.8 From OIL to DAML+OIL

DAML+OIL is the successor of OIL, defined in collaboration with research groups from the DARPA sponsored DAML programme, following the original versions of OIL and DAML-ONT (http://www.daml.org/2000/10/daml-ont.html).

2.8.1 Integration with RDFS

DAML+OIL is similar to OIL in many respects, but is more tightly integrated with RDFS, which provides the only specification of the language and its only serialization. While the dependence on RDFS has some advantages in terms of the re-use of existing RDFS infrastructure and the portability of DAML+OIL ontologies, using RDFS to completely define the structure of DAML+OIL is quite difficult as, unlike XML, RDFS is not designed for the precise specification of syntactic structure. For example, there is no way in RDFS to state that a restriction (slot constraint) should consist of exactly one property (slot) and one class. The solution to this problem adopted by DAML+OIL is to define the semantics of the language in such a way that they give a meaning to any (parts of) ontologies that conform to the RDFS specification, including 'strange' constructs such as slot constraints with multiple slots and classes. This is made easier by the fact that, unlike OIL, the semantics of DAML+OIL are directly defined in both a model theoretic and an axiomatic form (using KIF; Genesereth and Fikes, 1992). The meaning given to strange constructs may, however, include strange side effects. For example, in the case of a slot constraint with multiple slots and classes, the semantics interprets this in the same way as a conjunction of all the constraints that would result from taking the cross product of the specified slots and classes, but with the added (and possibly unexpected) effect that all these slot constraints must have the same interpretation (i.e. are equivalent). Although OIL's RDFS based syntax would seem to be susceptible to the same difficulties, in the case of OIL there does not seem to be an assumption that any ontology conforming to the RDFS meta-description would be a valid OIL ontology – presumably ontologies containing unexpected usages of the meta-properties would be rejected by OIL processors as the semantics do not specify how these could be translated into SHIQ(D).

DAML+OIL's dependence on RDFS also has consequences for the decidability of the language. In OIL, the language specification states that the slots used in cardinality constraints can only be applied to simple slots (slots that are neither transitive nor have transitive subslots). There is no way to capture this constraint in RDFS (although the language specification does include a warning about the problem), so DAML+OIL is theoretically undecidable. In practice, however, this may not be a very serious problem as it would be easy for a DAML+OIL processor to detect the occurrence of such a constraint and warn the user of the consequences.

Another effect of DAML+OIL's tight integration with RDFS is that the frame structure of OIL's syntax is much less evident: a DAML+OIL ontology is more DL- like in that it consists largely of a relatively unstructured collection of subsumption and equality axioms. This can make it more difficult to use DAML+OIL with frame based tools such as Protege (Grosso et al., 1999) or OilEd (Bechhofer et al., 2001b) because the axioms may be susceptible to many different frame-like groupings (Bechhofer et al., 2001a). From the point of view of language constructs, the differences between OIL and DAML+OIL are relatively trivial. Although there is some difference in keyword vocabulary, there is usually a one to one mapping of constructors, and in the cases where the constructors are not completely equivalent, simple translations are possible. For example, DAML+OIL restrictions (slot constraints) use `has-class` and `to-class` where OIL uses `ValueType` and `HasValue`, and while DAML+OIL has no direct equivalent to OIL's covering axioms, the same effects can be achieved using a combination of (disjoint) union and `subClass`. The similarities can clearly be seen in the DAML+OIL code below, which illustrates the DAML+OIL version of the OIL serialization in RDFS given above.

```
<daml:Ontology rdf:about="">
 <dc:title>"Swiss Life Skills DB</dc:title>
 <dc:creator>Ullrich Reimer</dc:creator>
 <dc:description>Part of the ontology from the Swiss
 Life case study</dc:description>
 <daml:versionInfo>1.0</daml:versionInfo>
</daml:Ontology>
<daml:Class rdf:about="#DesktopPublishing">
```

```
<rdfs:subClassOf>
  <daml:Class>
    <daml:intersectionOf rdf:parsetype=
        "daml:Collection">
      <daml:Class rdf:about="#Publishing"/>
      <daml:Class rdf:about=
        "#DocumentProcessing"/>
    </daml:intersectionOf>
  </daml:Class>
</rdfs:subClassOf>
</daml:Class>
<rdf:Description rdf:about=
  "#DesktopPublishingLevel3">
  <rdf:type rdf:about="#Skills"/>
  <SkillsLevel rdf:resource="#3"/>
</rdf:Description>
<daml:Class rdf:about="#DocumentProcessing">
  <rdfs:subClassOf rdf:about="#Skills"/>
</daml:Class>
<rdf:Description rdf:about="#GeorgeMiller">
  <rdf:type rdf:about="#Employee"/>
  <HasSkills rdf:resource=
    "#DesktopPublishingLevel3"/>
</rdf:Description>
<daml:ObjectProperty rdf:about="#HasSkills">
  <rdfs:domain rdf:about="#Employee"/>
  <rdfs:range rdf:about="#Skills"/>
</daml:ObjectProperty>
<daml:Class rdf:about="#ITProject">
  <rdfs:subClassOf rdf:about="#Project"/>
  <rdfs:subClassOf>
    <daml:Restriction>
      <daml:onProperty rdf:resource=
        "#ResponsibleDept"/>
      <daml:hasClass rdf:resource="#ITDept"/>
    </daml:Restriction>
  </rdfs:subClassOf>
</daml:Class>
</rdf:RDF>
```

2.8.2 Treatment of Individuals

The treatment of individuals in DAML+OIL is, however, very different from that in OIL. In the first place, DAML+OIL relies wholly on RDF for assertions on the type (class) of an individual or a relationship between a pair of individuals. In the second place, DAML+OIL treats individuals occurring in the ontology (in oneOf constructs or hasValue restrictions) as true individuals (i.e. interpreted as single elements in the domain of discourse) and not as primitive concepts as is the case in OIL. Moreover, there is no unique name assumption: in DAML+OIL it is possible to explicitly assert that two individuals are the same or different, or to leave their relationship unspecified. This treatment of individuals is very powerful, and justifies intuitive inferences that would not be valid for OIL, for example, that persons all of whose countries of residence are Italy are kinds of person that have at most one country of residence: Unfortunately, the combination of individuals with inverse roles is so powerful that no 'practical' decision procedure (for satisfiability/ subsumption) is currently known, and there is no implemented system that can provide sound and complete reasoning for the whole DAML+OIL language. In the absence of inverse roles, however, a tableaux algorithm has been devised (Horrocks and Sattler, 2001), and in the absence of individuals, DAML+OIL ontologies can exploit implemented DL systems via a translation into SHIQ similar to the one described for OIL. It would, of course, also be possible to translate DAML+OIL ontologies into SHIQ using the disjoint primitive concept of interpretation of individuals adopted by OIL, but in this case reasoning with individuals would not be sound and complete with respect to the semantics of the language.

2.8.3 DAML+OIL Data Types

The initial release of DAML+OIL did not include any specification of data types. However, in the March 2001 release, the language was extended with arbitrary data types from the XML Schema type system, which can be used in restrictions (slot constraints) and range constraints. As in SHOQ(D) (Horrocks and Sattler, 2001), a clean separation is maintained between instances of 'object' classes (defined using the ontology language) and instances of data types (defined using the XML Schema type system). In particular, it is assumed that that the domain of interpretation of object classes is disjoint from the domain of interpretation of data types, so that an instance of an object class (e.g. the individual Italy) can never have the same interpretation as a value of a data type (e.g. the integer 5), and that the set of object properties (which map individuals to individuals) is disjoint from the set of data type properties (which map individuals to data type values).

The disjointness of object and data type domains was motivated by both philosophical and pragmatic considerations:

- Data types are considered to be already sufficiently structured by the built-in predicates, and it is, therefore, not appropriate to form new classes of data type values using the ontology language (Hollunder and Baader, 1991).
- The simplicity and compactness of the ontology language are not compromised; even enumerating all the XML Schema data types would add greatly to its complexity, while adding a theory for each data type, even if it were possible, would lead to a language of monumental proportions.
- The semantic integrity of the language is not compromised; defining theories for all the XML Schema data types would be difficult or impossible without extending the language in directions whose semantics may be difficult to capture in the existing framework.
- The implementability of the language is not compromised; a hybrid reasoner can easily be implemented by combining a reasoner for the object language with one capable of deciding satisfiability questions with respect to conjunctions of (possibly negated) data types (Horrocks and Sattler, 2001).

From a theoretical point of view, this design means that the ontology language can specify constraints on data values, but as data values can never be instances of object classes, they cannot apply additional constraints to elements of the object domain. This allows the type system to be extended without having any impact on the object class (ontology) language, and vice versa. Similarly, reasoning components can be independently developed and trivially combined to give a hybrid reasoner whose properties are determined by those of the two components; in particular, the combined reasoner will be sound and complete if both components are sound and complete. From a practical point of view, DAML+OIL implementations can choose to support some or all of the XML schema data types. For supported data types, they can either implement their own type checker/validater or rely on some external component (non-supported data types could either be trapped as an error or ignored). The job of a type checker/validater is simply to take zero or more data values and one or more data types, and determine if there exists any data value that is equal to every one of the specified data values and is an instance of every one of the specified data types.

2.9 Experiences and Future Developments

Since its first release in December 2000, DAML+OIL has seen a remarkable take-up in the Semantic Web community. At leading recent Semantic Web events (e.g. the proceedings of the first Semantic Web Symposium, at http://www.semanticweb.org/swws/), many researchers are using DAML+ OIL for their work. The axiomatization (Fikes and McGuinness, 2001) has been formally verified (http://vis.home.mindspring.com/daml/kiflist.htm), languages are being built on top of it (Ankolenkar et al., 2002), it is being integrated with best-practice techniques in software engineering such as UML, and tutorials are being written for a wider audience (http://www.xml.com/pub/ a/2002/01/30/daml1.html).

Under the auspices of both the US DAML and the European IST programmes, a large number of tools have been written (http://www.daml.org/tools), including editors, storage, crawlers, visualizors, validators, inference engines, parsers, etc.

At the time of writing, the DAML ontology library contains over 175 ontologies, and DAML crawlers have found millions of DAML+OIL markup statements in documents. Possibly more important, however, is that some major efforts have been committed to encoding their ontologies in DAML+OIL. This has been particularly evident in the bio-ontology domain, where the Bio-Ontology Consortium has specified DAML+OIL as their ontology exchange language, and the Gene Ontology (The Gene Ontology Consortium, 2000) is being migrated to DAML+OIL in a project partially funded by GlaxoSmithKline Pharmaceuticals in cooperation with the Gene Ontology Consortium.

Large scale industrial take-up to date is still lacking. Perhaps this is too much to ask for, if we consider the very low industrial take-up of RDFS (a much simpler language than DAML+OIL), and, a bit more in the past, the long time that passed between the first XML specification and the first waves of industrial enthusiasm.

An important development is that the Semantic Web activity of the W3C has started a Web Ontology Working Group (http://www.w3.org/2001/sw/ webont/). This Working Group has been chartered to develop an ontology language for use in the Semantic Web, and DAML+OIL has specifically been designated by the charter to serve as the starting point for this development.

Thus, the most important impact of OIL and DAML+OIL may well not be the actual usage these languages get, but rather the fact that they form the basis of new languages, which will get widespread usage.

3

A Methodology for Ontology-based Knowledge Management

York Sure and Rudi Studer

3.1 Introduction

Ontologies are a core element of the knowledge management architecture described in Chapter 1. In this chapter we describe a methodology for application driven ontology development, covering the whole project lifecycle from the kick off phase to the maintenance phase. Existing methodologies and practical ontology development experiences have in common that they start from the identification of the purpose of the ontology and the need for domain knowledge acquisition. They differ in their foci and following steps to be taken. In our approach of the ontology development process, we integrate aspects from existing methodologies and lessons learned from practical experience (as described in the Section 3.7). We put ontology development into a wider organizational context by performing an a priori feasibility study. The feasibility study is based on CommonKADS (cf. Schreiber et al., 1999). We modified certain aspects of CommonKADS for a tight integration of the feasibility study into our methodology.

The case studies described in later chapters explore a broad spectrum of knowledge management challenges. Each has its own characteristics and usually evaluates only a subset of the generic methodology. The path of an application driven ontology development process is sketched in Figure 3.1. The main stream indicates activities that finally lead to a refined, evaluated

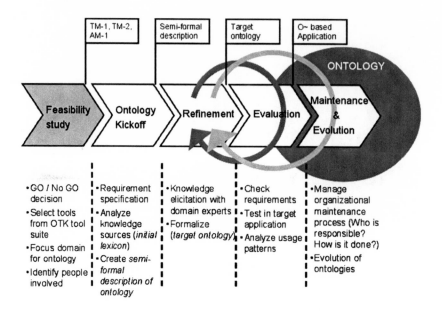

Figure 3.1   Steps of the on-to-knowledge (OTK) methodology

and applied ontology that has to be maintained. Each flag indicates major outcomes of the related activity. The most important steps of each activity are sketched below every activity. Refinement, evaluation and maintenance may need to be performed in iterative cycles. We now describe each activity in detail.

3.2 Feasibility Study

Any knowledge management system can function satisfactorily only if it is properly integrated into the organization in which it is operational. Many factors other than technology determine the success or failure of such a system. To analyse these factors, we initially have to prepare a *feasibility study*, that is, we identify problem/opportunity areas and potential solutions, and put them into a wider organizational perspective. In general, a feasibility study serves as decision support for economical, technical and project feasibility, in order to select the most promising focus area and target solution.

We focus on aspects of the feasibility study that help to identify (cf. the UML use case diagram (Booch et al., 1998) in Figure 3.2 for the following users and use cases and their interactions): (i) stakeholders related to a project

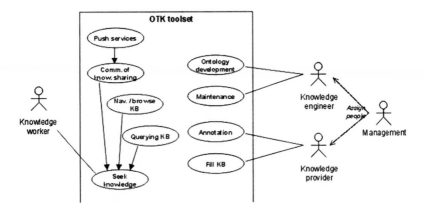

Figure 3.2   Users and use cases of on-to-knowledge

divided into *users of the system* (the knowledge worker) and *supporters of the system* (the knowledge engineer, the knowledge provider and management); (ii) use cases describing the usage scenarios which we call *user driven use cases* (push services, community of knowledge sharing, navigating and browsing a knowledge base, querying a knowledge base and seeking knowledge) and (iii) use cases supporting these user driven use cases which we call *supporting use cases* (ontology development, maintenance, annotation, fill knowledge base). The toolset described in later chapters is the underlying system that provides the technical backbone for instantiation of the use cases.

An essential element of a successful knowledge management system is the creation of a sound navigation system or framework to make it easy for users, viz. *knowledge workers*, to locate the knowledge they seek and to make it easy for users to provide the knowledge they want to share. The user can choose between two general approaches in finding information: one is in *querying the knowledge base*, and the other is in *navigating and browsing the knowledge base*. To support the provision and sharing of knowledge, the user has to be guided and supported in a *community of knowledge sharing*. *Push services* may push new knowledge according to user profiles to specific users that stated interest in particular knowledge areas.

Knowledge engineers provide the structure for the system, that is, ontologies. Initially they are responsible for the *ontology development*, but also for handling of the *maintenance* (and evolution) afterwards. *Knowledge providers* are responsible for the content provision to the system. Typically the main sources of that knowledge stem from intranet pages, documents in electronic formats and databases. The semantic interlinkage between intranets, electronic documents and a knowledge base are *annotations*. Web pages and documents have to be annotated to explicitly represent the semantics of their contents. The annotations might be stored within the sources or externally. Here we follow the approach of *filling* all annotations into the *knowledge base*. Last but not least, *management* has several important roles in a knowledge management project: it supports the legitimacy of the project and brings in vision that correlates with the overall company-wide vision. Management needs to be thoroughly convinced of the value of the project (Tiwana, 2000). Therefore it commits the resources needed – and especially assigns the people needed for the supporting use cases.

For the analysis of the users and use cases during the feasibility study, the CommonKADS methodology offers three models: the organization, task and agent model. The process of building these models proceeds in the following steps:

- Carry out a scoping and problem analysis study, consisting of two parts:

  a. identifying problem/opportunity areas and potential solutions, and putting them into a wider organizational perspective;
  b. deciding about economic, technical and project feasibility, in order to select the most promising focus area and target solution.

- Carry out an impacts and improvements study, for the selected target solution, again consisting of two parts:

  c. gathering insights into the interrelationships between the business task, actors involved, and use of knowledge for successful performance, and what improvements may be achieved here;
  d. deciding about organizational measures and task changes, in order to ensure organizational acceptance and integration of a knowledge system solution.

An overview of the process of organizational context modelling is given in Figure 3.3. Building the task, organization and agent model is done by following a series of steps supported by practical and easy-to-use worksheets and checklists (a detailed description of these steps is given in the CommonKADS methodology).

Following the feasibility study, the next step according to CommonKADS is to identify relevant tasks, agents carrying them out and knowledge items used by the agents while performing tasks.[1] For our purpose, the steps (viz. TM-1 worksheet, task analysis; TM-2 worksheet, knowledge item analysis;

---

[1] Some definitions from CommonKADS: 'A task is a piece of work that needs to be done by an agent.'; 'An agent is any human or software system able to execute a task in a certain domain.'

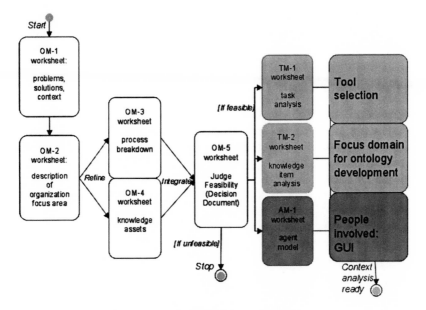

Figure 3.3   Modified CommonKADS steps

and AM-1 worksheet, agent model) lead to a modified result as indicated in the dark shading in Figure 3.3.

The *task analysis* (TM-1) identifies the features of relevant tasks and may therefore serve as a base for tool selection. To maximize the range of possible tool combinations, every candidate tool needs to be interoperable with respect to the architecture set out in Chapter 1.

The *knowledge item analysis* (TM-2) concentrates in detail on bottlenecks and improvements relating to specific areas of knowledge. It therefore helps to focus the domain for the ontology development and additionally serves as knowledge source for the kick off phase of the ontology development. Each identified knowledge item is a potential candidate for being part of the ontology and should be evaluated during the kick off phase.

The *agent model* (AM-1) also serves as an input source for the kick off phase of the ontology development by listing all relevant agents who possess knowledge items. Especially the human agents are likely to be potential domain experts and might be a valuable knowledge source during the kick off and refinement phase of the ontology development. Human agents might also be users of the system, that is, knowledge workers, and therefore this might indicate the design of an appropriate user interface for the application envisioned according to the needs of those agents.

Given that a GO decision was made during the feasibility study, the results as described above serve as input for the kick off phase of the ontology development.

3.3 Kick Off Phase

The ontology development starts with the kick off phase. Similar to software development in general, we start with an ontology requirements specification document (ORSD). In general it describes what an ontology should support, sketching the planned area of the ontology application and listing, for example, valuable knowledge sources. It contains a set of relevant structures of the domain (they are refined and formalized to a target ontology during the refinement phase). The ORSD should guide an ontology engineer in deciding about inclusion and exclusion of concepts/relations and the hierarchical structure of the ontology. In this early stage one should look for already developed and potentially reusable ontologies. In detail, the ORSD contains the following information:

1. *Domain and goal of the ontology.* First one should specify the particular domain in use (valuable input is given by the TM-2 worksheet), which might help to identify already existing ontologies. The feasibility study made clear proposals about interesting areas to be supported by a knowledge management project. The ontology engineer may use the outcomes of the task analysis to describe the goal of the ontology. The following list gives some examples: 'The ontology serves as a means to structure the *xy* domain', 'The ontology serves as a guideline for the knowledge distribution between department A and department B', 'Ontology serves as a base for semantic search.'.

2. *Design guidelines.* Design guidelines help users who are not familiar with modelling ontologies. They might, for example, contain an estimate of the number of concepts and the level of granularity of the planned model. This estimate is based on the knowledge item analysis, a further outcome of the feasibility study. For example, if the requirements analysis specified that an ontology should support browsing through a domain which includes around 100 concepts and the ontology engineer ended up with modelling 1000 concepts, either the ontology grew too big and should be modified to fulfil the requirements or the requirement specification is no longer up to date and should be updated. Also one might specify common rules how to name concepts. A typical approach for a naming convention is to begin all concepts with capitals and all relations with small caps. Whatever rules one might specify, they should be used consistently when modelling an ontology (e.g. supported by OntoEdit, cf. chapter 7).

3. *Knowledge sources.* The knowledge item analysis from the feasibility study serves as an important knowledge source at hand. The ontology engineer may derive people and documents to complete the list of knowledge sources for the domain in use. The following shows a partial list of knowledge sources as an example:

   - TM1 (based on CommonKADS);
   - domain experts (interviews, competency questionnaires);
   - (re-usable) ontologies;
   - dictionaries;
   - internal documents;
   - index lists;
   - regulations;
   - standard templates;
   - product and project descriptions;
   - technology white papers;
   - telephone indices;
   - web pages/site statistics;
   - organization charts;
   - employee role descriptions;
   - business plans;
   - external documents.

   The usage of potentially reusable ontologies may improve the speed and quality of the development during the whole process. These ontologies might, for example, give useful hints for modelling decisions. If the available ontologies fulfil the requirements (viz.in the ORSD) one might even reuse an already existing ontology – or reuse it with slight modifications. An ontology engineer should use all available knowledge sources based on their availability and reliability.

4. *(Potential) users and usage scenarios.* Lists of potential users or user groups and description of each usage scenario. These scenarios should be described from the potential user who may report from own experiences: In what situation did they wish such a system (better search for information, information distribution etc.)? How did they proceed without it? What were the hindering blocks? How would they like to be supported? The usage scenarios sketch the point of view of each individual user, which may vary to an extreme degree. Those views give interesting input to the structure of the ontology. The descriptions of the hindering blocks also include important hints for the design of the ontology based system. The acquisition of the usage scenarios is done via structured or informal interviews. A common way of modelling usage scenarios in software engineering are

use cases. In particular they help to identify stakeholders and to clarify their roles in the scenario.

5. *Competency questions.* The usage scenarios (see above) describe the real existing domain of the targeted system. They deliver information about concepts and their relations which have to be modelled in the target ontology. To derive that information out of the use cases, the ontology engineer has to transform the scenarios in detailed competency questions (Ushold and Grueninger, 1996). This represents an overview of possible queries to the system, indicating the scope and content of the domain ontology (cf. Chapter 7 for an example of a competency questionnaire).

6. *Applications supported by the ontology.* Here, typically a draft of the ontology based knowledge management application and its system and software environment is sketched. The ontology engineer may also use the task analysis from the feasibility study as an input source to describe the proposed system and analyse the role of the ontology. The draft must also deliver a clear picture about the ontology interface to the user and answer the following question: what parts of the ontology, namely concepts and relations, are visible to the user and how does he use them? If the application runs several times on different hosts, one might want to keep track of the different locations to enable separate update processes in the maintenance phase.

In general there are two concurrent approaches while modelling: top-down and bottom-up. The usage scenario/competency question method usually follows a *top-down* approach in modelling the domain. One starts by modelling concepts on a very generic level. Subsequently they are refined. This approach is typically done manually and leads to a high-quality engineered ontology. Available top-level ontologies may be reused here and serve as a starting point to develop new ontologies. In practice this seems to be more a *middle-out* approach, that is, to identify the most important concepts which will then be used to obtain the remainder of the hierarchy by generalization and specialization.

However, with the support of automatic document analysis, a typical *bottom-up* approach may be applied. There, relevant lexical entries are extracted semi-automatically from available documents. Based on the assumption that most concepts and conceptual structures of the domain as well the company terminology are described in documents, applying knowledge acquisition from text for ontology design seems to be promising. There exists also a bottom-up-driven approach for merging of ontologies. A number of proposals have been made to facilitate ontological engineering through automatic discovery from domain data, domain-specific natural language texts in particular (cf. Maedche and Staab, 2000). OntoExtract from CognIT (cf. Chapter 6) provides support for semi-automatic extraction of relevant concepts and relations between them.

Both approaches have advantages and drawbacks (e.g. the case studies at Swiss Life each follow a different path, cf. Chapter 12). The competency questions lead to a more detailed description of the problem area at hand. This supports the fine tuning of the ontology. On the other hand, this gathering of several views is likely to be never complete and might not focus on the documents available. Semi-automatic text extraction is usually not able to produce high-level quality but delivers instead a more complete list of relevant concepts. So, the top-down approach meets the representation of the information demand better than the bottom-up approach with automatic analysis of documents, which itself supports a better representation of the information supply. A promising method combines both approaches. We propose that ontology engineers should include various knowledge sources depending on their availability and their reliability (see above) and each time use the more applicable method to extract relevant knowledge from the sources.

3.4 Refinement Phase

The goal of the refinement phase is to produce a mature and application-oriented target ontology according to the specification given by the kick off phase. It is divided into several subphases. First, a knowledge elicitation process with domain experts based on the initial input from the kick off phase is performed. There, the initial draft of the ontology is modified and/or extended (e.g. by refining concepts). Second, the target ontology is created by formalizing the semi-formal description of the ontology in formal representation languages such as OIL and DAML+OIL (cf. Chapter 2).

Depending on the application that has to be supported, one has to choose the appropriate representation language. Formal representation languages typically differ in their expressive power and tool support for reasoning. The ontology engineer has to consider the advantages and limitations of the different languages to choose the appropriate one for the application.

This phase is closely linked to the evaluation phase. If the analysis of the ontology in the evaluation phase shows gaps or misconceptions, the ontology engineer takes these results as an input for the refinement phase. It might be necessary to perform several iterative steps.

3.5 Evaluation Phase

To describe the evaluation task, we cite Gomez-Perez (1996): 'to make a technical judgement of the ontologies, their associated software environment, and documentation with respect to a frame of reference ... The frame of reference may be requirements, specifications, competency questions, and/ or the real world.'

The ontology engineer needs to check, whether the target ontology itself suffices the ontology requirements specification document (cf. Section 4.2) and whether the ontology based application supports or answers the competency questions, analysed in the kick off phase of the project. Therefore the ontology is tested in the target application environment. A prototype should already show core functionalities of the target system. Feedback from beta users of the prototype may be a valuable input for further refinement of the ontology.

A valuable input for refinement (and further maintenance) are usage patterns of the ontology. The system has to track the ways users navigate or search for concepts and relations. With such an ontology log file analysis, one may trace what areas of the ontology are often used and others which were not navigated. Less frequently used parts of the ontology should be monitored whether they are relevant for the application. Parts of the ontology used with high frequency might need to be expanded. However, the ontology engineer should carefully evaluate the usage patterns before updating the ontology.

3.6 Maintenance and Evolution Phase

The maintenance and evolution of an ontology-based application is primarily an organizational process. There have to be strict rules to the update/insert/ delete processes of ontologies. We recommend that the ontology engineer gathers changes to the ontology and initiates the switch over to a new version of the ontology after thoroughly testing all possible effects on the application. Most important is to clarify who is responsible for maintenance and how it is performed. For example, is a single person or a consortium responsible for the maintenance process? In which time interval is the ontology maintained?

Seminal proposals for handling of the evolution process of ontologies already exist. See Chapter 4 for a detailed description of these aspects.

3.7 Related Work

Methodologies that guide the building process of ontologies have been proposed by several research groups. Due to the fact that ontology engineering is still a relatively immature discipline, each research group employed its own methodology. We have based our methodology on previously published methodologies.

In recent years, some research groups have proposed methodologies guiding the ontology development process. Ushold's skeletal methodology was the first methodological outline proposed in 1995 on the basis of the experience gathered in developing the Enterprise Ontology (Ushold and King, 1995). On the basis of the Toronto Virtual Enterprise (TOVE) project, Ushold and Grueninger (1996) described ontology development steps. A method to build an ontology in the domain of electrical networks was presented from Bernaras et al. (1996) as part of the Esprit KACTUS project. At the same time Methontology appeared (Gomez-Perez, 1996), extended in later papers. In parallel, the philosophical discipline of ontology is evolving towards an engineering discipline. Guarino and Welty (2000) demonstrate how some methodology efforts founded on analytic notions that have been drawn from philosophy can be used as formal tools of ontological analysis. In the following, we give a brief overview of these methodologies.

3.7.1 Skeletal Methodology

This methodology is based on the experience of building the enterprise ontology (Ushold and King, 1995), which includes a set of ontologies for enterprise modelling. The following guidelines for developing ontologies are proposed:

1. *Identify purpose.* Clarify goal and intended usage of the ontology.
2. *Building the ontology*, which is broken down into three steps:

   a. *Ontology capture.* Identify key concepts and relationships in the domain of interest. Create precise unambiguous text definitions for such concepts and relationships and identify terms to refer to them. Use a middle-out approach to perform this step, so identify the most important concepts which will then be used to obtain the remainder of the hierarchy by generalization and specialization.
   b. *Coding.* Represent the knowledge acquired in 2(a) in a formal language.
   c. *Integrate* existing ontologies.

3. *Evaluation.* Make a judgement of the ontologies with respect to a frame of reference which may be requirements specifications or competency questions.
4. *Documentation.* Document ontologies according to the type and purpose.

A disadvantage of this methodology is that it does not precisely describe the techniques for performing the different activities. For example, it remains unclear how the key concepts and relationships should be acquired, only a very vague guideline, involving the use of brainstorming techniques, is given

(López, 1999). A life cycle is not recommended. There are no guidelines about the maintenance of evolving ontologies.

We catch up with the idea of competency questions and expand their usage. We not only propose to use them for evaluation of the system, but also for finding relevant lexical entries such as concepts, relations, etc. (see Chapter 8 for an example).

3.7.2 KACTUS

The approach of Bernaras et al. (1996) was developed within the Esprit KACTUS project. One of the objectives of this project was to investigate the feasibility of knowledge re-use in complex technical systems and the role of ontologies to support it. The methodology recommends an application driven development of ontologies. So, every time an application is assembled, the ontology that represents the knowledge required for the application is built. Three steps have to be taken every time an ontology-based application is assembled:

1. Specification of the application. Provide an application context and a view of the components that the application tries to model.
2. Preliminary design. Based on relevant top-level ontological categories, create a first draft where the list of terms and application specific tasks developed during the previous phase is used as input for obtaining several views of the global model in accordance with the top-level ontological categories determined. Search for existing ontologies which may be refined and extended for use in the new application.
3. Ontology refinement and structuring. Structure and refine the model in order to arrive at a definitive design.

The methodology offers very little detail and does not recommend particular techniques to support the development steps. Also, documentation, evaluation and maintenance processes are missing (López, 1999). In general we agree with the general idea of application driven ontology development and in particular with refinement and structuring, which is reflected by our proposal for the ontology development process.

3.7.3 Methontology

The methontology framework from Gomez-Perez (1996) includes:

1. The identification of the ontology development process, which refers to which tasks (planning, control, specification, knowledge acquisition,

conceptualization, integration, implementation, evaluation, documentation, configuration management) one should carry out, when building
ontologies.

2. The identification of stages through which an ontology passes during its
lifetime.

3. The steps to be taken to perform each activity, supporting techniques and
evaluation steps.

4. Setting up an ORSD to capture requirements for an ontology similar to a
software specification.

The methodology offers detailed support in development-oriented activities
except formalization and maintenance, and describes project management
activities. We used the methontology framework as a skeleton and tailored
it for our specific need. We adopted the idea of an ORSD, but modified and
extended the version presented for our own needs. For example, we included
design guidelines to support ontology engineers and especially to ensure a
consistent development of ontologies. Besides typical modelling guidelines,
they also include constraints that should be checked during the development
phase.

3.7.4 Formal Tools of Ontological Analysis

In Guarino and Welty (2000), a formal ontology of unary properties is
proposed to help the use of the subsumption relation in a disciplined way.
This formal ontology is based on four fundamental philosophical notions
(identity, unity, rigidity and dependence) which impose constraints about
the way subsumption is used to model a domain. These constraints clarify
misconceptions about taxonomies and give support to bring substantial order
to ontologies. The authors present a six-step methodology to produce well-
founded taxonomies. These steps are important to get a cleaner taxonomy due
to the semantic constraints imposed on the is-a relation. In addition, the
rigorous analysis forces the ontology engineer to make ontological commitments explicit, clarifying the intended meaning of the concepts used and
therefore producing a more re-usable taxonomic structure. We take this into
account in the refinement phase of the application driven ontology process.

3.8 Conclusion

We have presented a comprehensive methodology that guides the development of ontologies for knowledge management applications. Five major steps
– a feasibility study, kick off phase, refinement phase, evaluation phase and

maintenance and evolution phase – are performed to build an ontology-based application. In the future we plan to expand in particular support for the maintenance and evolutionary aspects of ontologies, which is still a wide open research area.

4

Ontology Management: Storing, Aligning and Maintaining Ontologies

Michel Klein, Ying Ding, Dieter Fensel and Borys Omelayenko

4.1 The Requirement for Ontology Management

Ontologies need to be stored, sometimes aligned and their evolution needs to be managed. All these tasks together are called *ontology management*.

Alignment is a central task in ontology re-use. Re-use of existing ontologies often requires considerable effort (Uschold et al., 1998): the ontologies either need to be *integrated* (Pinto et al., 1999), which means that they are merged into one new ontology, or the ontologies can be kept separate. In both cases, the ontologies have to be aligned, which means that they have to be brought into mutual agreement. The problems that underlie the difficulties in integrating and aligning are the *mismatches* that may exist between separate ontologies. Ontologies can differ at the language level, which can mean that they are represented in a different syntax, or that the expressiveness of the ontology language is dissimilar. Ontologies also can have mismatches at the model level, for example, in the paradigm, or modelling style (Klein, 2001).

Ontology alignment is very relevant in a Semantic Web context. The Semantic Web will provide us with a lot of freely accessible domain specific ontologies. To form a real *web of semantics* – which will allow computers to combine and infer implicit knowledge – those separate ontologies should be aligned and linked.

Support for evolving ontologies is required in almost all situations where ontologies are used in real-world applications. In those cases, ontologies are often developed by several persons and will continue to evolve over time, because of changes in the real world, adaptations to different tasks, or alignments to other ontologies. To prevent that such changes will invalidate existing usage, a change management methodology is needed. This involves advanced versioning methods for the development and the maintenance of ontologies, but also configuration management, that takes care of the identification, relations and interpretation of ontology versions.

All these aspects come together in integrated *ontology library systems*. When the number of different ontologies is increasing, the task of storing, maintaining and re-organizing them to secure the successful re-use of ontologies is challenging. Ontology library systems can help in the grouping and re-organizing ontologies for further re-use, integration, maintenance, mapping and versioning. Basically, a library system offers various functions for managing, adapting and standardizing groups of ontologies. Such integrated systems are a requirement for the Semantic Web to grow further and scale up.

In this chapter, we describe a number of results with respect to the above mentioned areas. We start with a description of the alignment task and show a meta-ontology that is developed to specify the mappings. Then, we discuss the problems that are caused by evolving ontologies and describe two important elements of a change management methodology. Finally, in Section 4.4 we survey existing library systems and formulate a wish-list of features of an ontology library system.

4.2 Aligning Ontologies

For effective ontology interoperation, ontologies must be efficiently aligned. These alignments must explicitly represent the maximal possible share of the relationships between the ontologies and their elements to enable efficient ontology re-use.

4.2.1 Why is Aligning Needed

The knowledge management scenario, which is in focus in this volume, assumes different departments and individual employees create domain-specific ontologies capturing specific aspects of their knowledge. Special mapping ontologies must be created to link different terminologies and modelling styles used in these domain specific ontologies, creating bridges between separated pieces of knowledge. These bridges along with domain ontologies are then used to perform cross-ontology information search and retrieval.

Existing ontology mapping techniques primarily concern the weak ontology coupling (Mitra et al., 2000) needed to refer them and to query. Inference-based techniques can be also used to represent and exploit the correspondences between ontology terms (Sintek and Decker, 2001). However, practical knowledge integration and aligning tasks require development of computationally cheap mapping techniques.

4.2.2 Aligning Annotated XML Documents

Companies tend to use XML to represent more and more internal documents. One of the main benefits of using XML is the possibility of automatic processing of XML documents according to their conceptual models represented with document DTDs or XML schemas. The documents may contain knowledge-level annotations represented in RDF according to a certain ontology encoded in RDF Schema. These annotations may be used to answer knowledge-level user's queries and to perform document search and retrieval based on the results of processing the annotations and ontologies. Query results may themselves be represented as XML documents to be further automatically processed.

This knowledge management task requires performing a sequence of transformations of the concepts represented in the documents illustrated in Figure 4.1.

- The source XML document is transformed to its conceptual model in RDF (schema) that captures all the objects (with a possible shallow hierarchy of their classes) and relations presented in the document. XML DTDs and schemas contain a large portion of knowledge about the concepts repre-

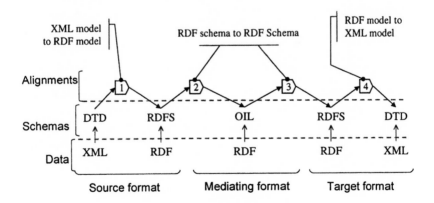

Figure 4.1   The alignments

sented in the documents. Basically, most of the part-of relations are explicitly encoded and can be re-used in document conceptual models. Some efficient techniques for doing that have been already proposed (Lee and Chu, 2001; Mello and Heuser, 2001). Provided with the results of these algorithms, the user needs only to align the models to some upper-level ontologies and perform small modifications.

- The source conceptual model is aligned to the mediating ontology provided by the company and, for example, classifying all the documents according to a taxonomy of possible subjects. The document is then stored as an instance of the mediating ontology. The query service needs to process the conceptual models and the mediating ontology to extract the concepts that match the query.

- The target RDF (schema) conceptual model corresponds to the target document and is aligned to the mediating ontology. The target document is extracted from the mediating document according to these alignments.

- The target XML document serialization is re-constructed from its conceptual model to be then processed by the querying service.

A mapping meta-ontology needs to be developed to represent all the bridges between XML elements and attributes, and RDF Schema classes and properties.

The XML documents tend to change with time and it may not be efficient to store all of them as RDF instances of their conceptual models. Instead, the documents that were retrieved to answer the query must be transformed into the format required for the query answer. The transformation of instance XML documents may be performed by an XSLT (Clark, 1999) stylesheet specially compiled from the mappings to transform the concepts from the source documents to the XML format required by the query source. The necessity to compile the mappings naturally restricts the expressive power of the mapping language.

4.2.3 Mapping Meta-ontology

We developed the RDF transformation (RDFT) mapping meta-ontology[1] that specifies a small ontology for mapping XML DTDs to/and RDF Schemas and is built on top of RDF Schema. The basic class diagram is presented in Figure 4.2, where the classes are represented by their names, and name nesting indicates the is-a relationship. The main concept of RDFT is the bridge between two sets of concepts: the source set and the target set. The bridges are grouped into maps. Each **Map** is a collection of bridges serving a single purpose. The

---

[1] http://www.cs.vu.nl/~borys/rdft

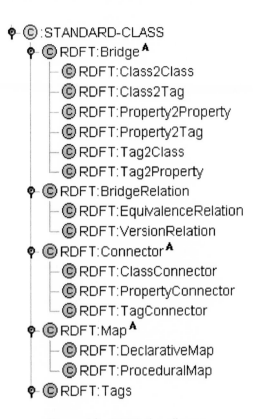

Figure 4.2   RDFT class diagram

maps are identified by their names (URIs) and form minimal re-usable modules of RDFT bridges.

An abstract class **Bridge** describes common properties of bridges allowing only one-to-many and many-to-one bridges. Each `Bridge` contains the `ValueCorrespondence` property linking a map between the instance values of the source and target entities.

The bridges also contain the `Relation` property linking to one of the `BridgeRelations`: `EquivalenceRelation` or `VersionRelation`:

- `Equivalence` bridges specify that the source element of a one-to-many bridge is equivalent to the target set of elements, and the source set of elements is equivalent to the target element for many-to-one bridges.
- A `Version` bridge specifies that the target set of elements form a (later)

version of the source set of elements. Opposite to equivalence bridges, they assume that both source and target concepts belong to the same domain (or document standard), and may refer to two concepts with the same name (but different namespaces indicating versions), and imply that all the relations that held for the original concept must hold for the versioned concept, if the opposite is not stated explicitly.

Several types of Bridges are defined in RDFT:

• Class2Class and Property2Property bridges between RDF Schema classes and properties. In RDF Schema, classes are represented by their names, place in taxonomy, and properties that are attached to this class. Properties are defined as first-class objects together with classes, and they capture most of domain knowledge (Lassila and Swick, 1999). Classes specify aggregation of properties, and thus we do not include class-to-property and property-to-class bridges in RDFT. These bridges occur at steps 2 and 3 of the integration process depicted in Figure 4.1. Class2Class bridges between a set of *n* source classes and a set of *m* target classes declares that a set of *n* instances of the classes listed as sources corresponds to a set of *m* instances of the classes listed as targets. Property2Property bridges have similar semantics related to properties.

• Tag2Class and Tag2Property bridges between XML tags of the source DTD and the target RDF Schema classes and properties. They occur at step 1 of the integration process.

• Class2Tag and Property2Tag bridges between RDF Schema classes and properties, and the elements of the target DTD. They occur at step 4 of the process.

All of the bridges contain the ValueCorrespondence property inherited from the abstract Bridge class linking to a map. Two types of Maps are defined in RDFT:

• DeclarativeMap specifies a set of bridges mapping all possible values to be mapped.

• ProceduralMap specifies an XPath (Clark, 1999) expression transforming instance data. XPath defines the means for two tasks: addressing data elements in XML documents and performing element or attribute value transformations (Chapter 4 of the specification[2]). In procedural maps we use only the second part of the XPath functions (e.g. substring_ before).

---

[2] http://www.w3.org/tr/xpath

The bridges are linked to the maps with different `Connectors`. Each `Connector` represents a class-property assignment that has a unique URI and as a result can be distinguished from other assignments of a property to a class.

More information on RDFT is available from the RDFT project homepage.[3]

4.2.4 Mapping in OIL

The RDFT meta-ontology is intended to serve as a template for creating mapping ontologies for the business integration tasks. It represents meta-classes that are then instantiated into user's classes linking user's domain ontologies. OIL was developed as an ontology representation language and contains less flexible means for defining meta-classes than RDF Schema does. However, it is still possible to represent RDFT semantics in OIL.

We model the RDFT bridges in OIL in the following way. For each bridge, we specify two classes that define the source and the target concepts to be mapped. For example, a one-to-many `Class2Class` bridge specifies the fact that each instance of the source class is equivalent to the set of instances, one instance of each target class. The corresponding bridge sources and bridge targets can be defined as follows:

```
class-def BridgeSources
   slot-constraint rdft_set_member has-value Source-
   Class
```

```
class-def BridgeTargets
   slot-constraint rdft_set_member has-value Target-
   Class1
   slot-constraint rdft_set_member has-value Target-
   Class2
   slot-constraint rdft_set_member has-value Target-
   Class3
```

The equivalence of `BridgeSources` to `BridgeTargets` can also be modelled in OIL by an `equivalent` axiom:

```
equivalent BridgeSources BridgeTargets
```

This axiom specifies the fact that each instance of the `SourceClass` class is equivalent to three instances, one of the `TargetClass1`, the second is of the `TargetClass2`, and the third is of `TargetClass3`.

However, statements over the bridges (e.g. `Maps`) need to be specified as

---

[3] http://www.cs.vu.nl/~borys/rdft

statements over OIL axioms, and it is problematic to represent them in OIL directly. Despite that, bridge translation to OIL is still very useful because it allows us to invoke an inference engine to perform knowledge-level validation of the bridges.

4.3 Supporting Ontology Change

4.3.1 Ontologies are Changing

In practice, ontologies are not static, but evolve over time. Support to handle this evolution is needed. This is especially important when ontologies will be used in a decentralized and uncontrolled environment like the Web, where changes occur without coordination. Much more than in a controlled environment, this may have unexpected and unknown results.

There are several reasons for changes in ontologies. According to Gruber (1993), an ontology is a *specification of a conceptualization of a domain*. Hence, changes in ontologies can be caused by either:

- changes in the domain;
- changes in the conceptualization;
- changes in the specification.

The first type of change occurs often. This problem is very well known from the area of database schema versioning. In Ventrone and Heiler (1991), seven different situations are sketched in which changes in a domain (domain evolution) require changes to a database model. An example of this type of change is the merging of two university departments: this is a change in the real world, which requires the ontology that describes this domain to be modified too.

Changes in the conceptualization are also happening frequently. It is important to realize that a *shared* conceptualization of a domain – which is a requirement for information exchange – is not a static specification that is produced once in the history, but has to be reached over time. In chapter 14, ontologies are described as dynamic networks of meaning, in which consensus is achieved in a social process of exchanging information and meaning. This view attributes a dual role to ontologies in information exchange: they provide consensus that is both a *prerequisite* for information exchange and a *result* of this exchange process.

A conceptualization can also change because of the usage perspective. Different tasks may imply different views on the domain and consequently a different conceptualization. When an ontology is adapted for a new task or a new domain, the modifications represent changes to the conceptualization. For example, consider an ontology about traffic connections in Amsterdam, with concepts like roads, cycle-tracks, canals, bridges and so on. When the ontology is adapted from a bicycle perspective to a water transport perspective, the conceptualization of a bridge changes from a remedy for crossing a canal to a time consuming obstacle.

Finally, a specification change is a kind of translation, that is, a change in the way in which a conceptualization is formally recorded. Although ontology translation is an important and non-trivial issue in many practical applications, it is less interesting *from a change management perspective*, for two reasons. First, an important goal of a translation is to retain the semantics, that is, specification variants should be equivalent[4] and they thus only cause syntactic interoperability problems. Second, a translation is often created to use the ontology in an other context (i.e. an other application or system), which heavily reduces the importance of interoperability questions.

Changes in ontologies are thus inevitable. In the next sections, we look at the characteristics of ontology changes and describe two elements of an ontology change methodology.

4.3.2 Changes in Ontologies Involve Several Problems

There are several problems involved with ontology changes. In this section we look at incompatibilities caused by ontology changes, the specification of them, and at the conceptual implication of such changes.

4.3.2.1 Effects of Ontology Change

An important effect of the evolution of ontologies is that it might cause incompatibilities. Incompatibility for ontologies means that the original ontology cannot be replaced by the changed version without causing side effects in the conforming data or the applications that use them. However, the real problem is that these side effects, and thus the meaning of compatibility, depend on the *use* of the ontology.

- When an ontology is used to specify the meaning of data, this data may get a different interpretation or may use unknown terms. An example of this use is a web page in which content is annotated with terms from an ontology.
- If ontologies are built from other ontologies, changes to the source ontology may affect the meaning of the resulting ontologies.

---

[4] Although in practice a translation often implies a change in semantics, possibly caused by differences in the representation languages. See Corcho et al. (2000) and Klein (2001) for a discussion of ontology language differences and mismatches.

- Applications that use the ontology may also be hampered by changes to the ontology. In the ideal case, the conceptual knowledge that is necessary for an application should be merely specified in the ontology; however, in practice applications also use an internal model. This internal model may become incompatible with the ontology.

The meaning of compatibility is different for each of those types of usage. In the first case, compatibility means the ability to interpret all the data correctly through the changed ontology. This is much like the interpretation of compatibility in database schema versioning. Compatibility here means 'preservation of instance data'.

In the second case, the effects of the changes on the logical model that the ontology forms are often important. Other ontologies that import an ontology might depend on the conclusions that can be drawn from it. A change in the ontology should not make previous conclusions invalid. In this case, compatibility means 'consequence preservation'.

Applications that use the ontology might depend on the logical model, but also on the characteristics of the ontology itself. For example, a web site that uses an ontology for navigation can depend on the fact that there are only four top-level classes, or that the hierarchy is only three levels deep. A change that does not invalidate queries to instance data or the logical model might invalidate queries to the ontology itself. This interpretation of compatibility is 'preservation of answers to ontology queries'.

4.3.2.2 Typical Changes and their Specification

The specification of changes is another problem. There are many possible types of changes in ontologies, ranging from simple renamings to compound transformations. The specification of especially the latter is important, because the effect of a compound change can be different from the accumulated effect of steps that build the complex change (Lerner, 2000).

To make this more concrete, we consider changes in a particular content standard, that is, UNSPSC.[5] Content standards specify a standard hierarchy of products and services which can be used by companies to classify their actual products. This hierarchy can be considered as a simple ontology that specifies a consensus on the products that exist. Different companies that use the same content standard can easily communicate with respect to their products. Besides UNSPSC, which addresses a general and broad domain of products and services, there are several other standard classifications in use, for exam-

---

[5] http://eccma.org/unspsc/

ple, RosettaNet,[6] which is targeted at the IT industry, and e@Class,[7] another broad standard that originates from Germany.

These standards tend to change very often. For example, when we take a look at UNSPSC, we see the following:

- there were 16 updates between 31 January 2001 and 14 September 2001;
- each update contained between 50 and 600 changes;
- in 7.5 months, more than 20% of the current standard is changed!

Although some parts of the UNSPSC schema might be more stable than other parts, it is clear that this amount of changes cannot be ignored. Such a high change rate can quickly invalidate a lot of the actual classifications of products. For example, the product 'Binding elements' in version 8.0 is removed from the standard and three new products are added in version 8.1 ('Binding spines or snaps', 'Binding coils or wire loops', and 'Binding combs or strips'). This means that all products that were classified as 'Binding elements' are unclassified under the new version.

An analysis of differences between two version of content standards has yielded the following list of typical changes: class-title changes, additions of classes, relocations of classes in the hierarchy (by moving them up or down in the hierarchy, or horizontally), relocations of a whole subtree in the hierarchy, merges of two classes (in two variants: two classes become one new class, or one class is appended to the other class), splits of classes, and pure deletions. However, current versioning techniques for content standards are often quite simple. In UNSPSC, for example, all changes are encoded as either additions, deletions or edits (title changes). This means that the relocation of a subtree is specified as a sequence of 'delete a list of classes' and 'add a list of classes'.

4.3.2.3 Conceptual Implication of Changes

Another problem with ontology change is the possible discrepancy between changes in the specification and changes in the conceptualization. The actual specification of concepts and properties is a *specific* representation of the conceptualization; however, the same concepts could also be specified differently. Hence, a change in the specification does not necessarily coincide with a change in the conceptualization (Klein and Fensel, 2001), and changes in the specification of an ontology are not by definition ontological changes.

---

[6] http://www.rosettanet.org/
[7] http://www.eclass.de/

For example, there are changes in the definition of a concept which are not meant to change the concept, and, the other way around, a concept can change without a change in its logical definition. An example of the first case is attaching a slot 'fuel-type' to a class 'Car'. Both class definitions still refer to the same ontological concept, but in the second version it is described more extensively. On the other hand, a natural language definition of a concept might change without a logical change in the definition of a concept, for example, a new definition of 'Chair' might exclude reclining chair.

In the literature, these different types of changes are distinguished in the following way (Visser et al., 1997b; Klein, 2001):

- a *conceptual change* is a change in the interpretation of a domain (i.e. the conceptualization), which results in different ontological concepts or different relations between those concepts;
- an *explication change* is a change in the way the conceptualization is specified.

It is impossible to determine the type of change automatically, because this is basically a decision of the ontology engineer. Therefore, it is necessary to allow ontology engineers to specify the intention of their change. If they characterize a change in a definition as 'conceptual', then the source and target definitions should be considered as different (even when their specification is the same), else, if a change is 'explicational', the two definitions can be regarded as equivalent.

4.3.3 Change Management

A change management methodology that allows partly automatic transformation of data and ontologies between different versions is essential. Such a methodology should be able to cope with the different types of incompatibility, should allow a precise specification of changes, and should help ontology engineers to specify the conceptual consequence of the change. We now discuss two aspects of such a methodology: a comparison tool for ontologies and the change specification mechanism.

4.3.3.1 Comparing Ontologies

An important aspect of a change management methodology is the ability to compare versions of ontologies and highlight the differences. This helps in finding changes in ontologies, even if those have occurred in an uncontrolled way, that is, possibly by different people in an unknown order. We have developed an Ontology Versioning Server, a web-based system to manage changes in ontologies. Its main function is to present a transparent interface to arbitrary versions of ontologies. To achieve this, the system maintains an internal specification of the relation between the different variants of ontologies. It allows users to differentiate between ontologies at a conceptual level and to export the differences as adaptations or transformations.

One of the central features of system is the ability to compare ontologies at a conceptual level. This is inspired by UNIX `diff`, but the implementation is quite different. Standard `diff` compares file versions at line level, highlighting the lines that textually differ in two versions. Our system, in contrast, compares version of ontologies at a *structural* level, showing which definitions of ontological concepts or properties are changed.

The comparison function distinguishes between the following types of change:

- Non-logical change, e.g. in a natural language description. These are changes in the label of an concept or property, or in comment inside definitions.
- Logical definition change. This is a change in the definition of a concept that affects its formal semantics. Examples of such changes are alterations of subclass statements, or changes in the domain or range of properties. Additions or deletions of local property restrictions in a class are also logical changes. The second and third change in Figure 4.3 (class 'Male' and property 'hasParent') are examples of such changes.
- Identifier change. This is the case where a concept or property is given a new identifier, that is, a renaming.
- Addition of definitions.
- Deletion of definitions.

Each type of change is highlighted in a different colour, and the actually changed lines are printed in boldface. An example of the visual representation of the result of a comparison is shown in Figure 4.3.

The comparison function also allows the user to *characterize* the conceptual implication of the changes. For the first three types of changes, the user is given the option to label them either as 'identical' (i.e. the change is an explication change), or as 'conceptual change'. In the latter case, the user can specify the conceptual relation between the two version of the concept. For example, by stating that the property 'hasParent$_{1.0}$' is a sub-property of 'hasParent$_{2.0}$'.

Another function is the possibility of analysing effects of changes. Changes in ontologies not only affect the data and applications that use them, but they can also have unintended, unexpected and unforeseeable consequences in the ontology itself (McGuinness et al., 2000). The system provides some basic

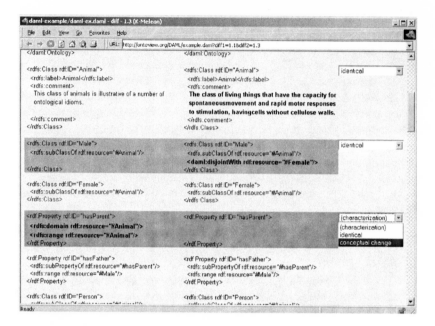

Figure 4.3 The result of a comparison of two ontologies

support for the analysis of these effects. First, on request it can highlight the places in the ontology where conceptually changed concepts or properties are used. For example, if a property 'hasChild' is changed, it will highlight the definition of the class 'Mother', which uses the property 'hasChild'. This function can also exploit the transitivity of properties to show the propagation of possible changes through the ontology.

4.3.3.2 Specification of Change

A change in an ontology constitutes a new version of the ontology. This new version defines an orthogonal *update* relation between the definitions in the original version of the ontology and those in the new version. The update relation between two versions of a concept, for example, between class $A_{1.0}$ and class $A_{2.0}$, is fundamentally different from the relation between two concepts inside an ontology, for example, between class A and class B. In the latter case, the relation is purely conceptual; however, the update relation also has meta-information about the change of the concept associated with it.

We distinguish the following properties that are associated with an update relation:

- *transformation* or *actual change*: a specification of what has actually changed in an ontological definition, specified by a set of change operations (cf. (Banerjee et al., 1987), for example, change of a restriction on a property, addition of a class, removal of a property, etc.;
- *conceptual relation*: the logical relation between constructs in the two versions of the ontology, for example, specified by equivalence relations, subsumption relations, logical rules, or approximations. The conceptual relation between two versions of a concept specifies the intention of the ontology engineer that characterized the change.
- descriptive meta-data like *date*, *author*, and *reason* of the update: this describes the when, who and why of the change;
- *valid context*: a description of the context in which the update is valid. In its simplest form, this is the time period in which the change is valid in the real world, conforming to *valid date* in temporal databases (Roddick, 1995) (in this terminology, the 'date' in the descriptive meta-data is called *transaction date*). More extensive descriptions of the context, in various degrees of formality, are also possible.

Keeping track of these four aspects of a change relation serves several functions. It is possible to perform loss-less transformations of ontologies, by exploiting the set of change operations. The conceptual relation gives the ability to re-interpret data and other ontologies that use the changed ontology via the new ontology. The meta-data and context helps to select versions and validate their applicability.

4.4 Organizing Ontologies

As the number of different ontologies is increasing, the task of storing, maintaining and re-organizing them to ensure the successful re-use of ontologies is challenging (Fensel, 2001). *Ontology library systems* are an important tool in grouping and re-organizing ontologies for further re-use, integration, maintenance, mapping and versioning. Basically it is a library system that offers various functions for managing, adapting and standardizing groups of ontologies. It should be easily accessible and offer efficient support for re-using existing relevant ontologies and standardizing them based on upper-level ontologies and ontology representation languages (Ding and Fensel, 2001).

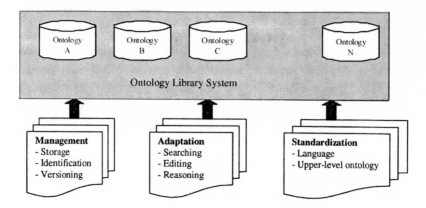

Figure 4.4   Aspects of an ontology library system

4.4.1 Sesame Requirements

Sesame[8] (see Chapter 5) allows persistent storage of RDF data and RDFS information and subsequent querying of the information enabled by RQL. Sesame selected a relational database as the storage mechanism, but is DBMS-independent via the repository abstraction layer (RAL). The RAL is an interface that offers RDF-specific methods to its clients and translates these methods to calls to its specific DBMS. The big advantage of RAL is that it makes it possible to implement Sesame on top of a wide variety of repositories without changing any of Sesame's other components (Broekstra et al., 2000).

Sesame provides a basis functionality to store ontologies and their instances and provides the querying service as well. However, for large-scale use and ontology re-use, more advanced functions are needed. A real-world ontology library system must support the following (see Figure 4.4):

- open storage, identification and versioning of ontologies;
- smooth *access* to existing ontologies and advanced support in *adapting* ontologies to certain domain and task-specific circumstances (instead of requiring such ontologies to be developed from scratch);
- fully employing the power of *standardization* and providing access to *upper-layer ontologies* and standard *representation languages.*

4.4.2 Functionality of an Ontology Storage System

An ontology library system should feature a functional infrastructure to store and maintain ontologies, an uncomplicated adapting environment for editing,

[8] http://sesame.aidministrator.nl/

searching and reasoning ontologies, and strong standardization support by providing upper-level ontologies and standard ontology representation languages. The aspects above can be further specified as follows.

4.4.2.1 Management

The main purpose of ontologies is to enable knowledge sharing and re-use (Visser et al., 1997a). Important functions should include open storage, identification, and versioning support.

- *Storage* (how to store the ontology): (a) Is the ontology easily accessible (via a client/server architecture, Peer-to-Peer, etc.); (b) Are ontologies classified according to some existing or home-made categories; and (c) Are ontologies stored in modules? (The modularity structure can guarantees proficient ontology re-use).
- *Identification* (how to uniquely identify an ontology): Each ontology must have a unique identifier in the ontology library system.
- *Versioning* (how to maintain the changes of ontologies in an ontology library system): Versioning is very critical in ensuring consistency among different versions of ontologies.

4.4.2.1 Adaptation

Ontology library systems should facilitate the task of extending and updating ontologies. They should provide user-friendly environments for searching, editing and reasoning ontologies. Important aspects include support in finding and modifying existing ontologies.

- *Searching* (how to search ontology): Does a library system provide certain searching facilities, such as keyword-based searching or other advanced searching? Does it feature an adequate browsing function?
- *Editing* (how to add, delete and edit specific ontologies): How does the system support the editing function? Does it support remote and cooperative editing?
- *Reasoning* (how to derive consequences from an ontology): How does the system support ontology evaluation and verification? Is it possible to derive any query-answering behaviour?

4.4.2.3 Standardization

Ontology library systems should support existing or available standards, such as standardized ontology representation languages and standardized taxonomies or structures of ontologies.

- *Language* (the kind of standard ontology language used in the ontology library system, for instance, RDFS,[9] XMLS[10] or DAML+OIL[11]): Does the system only support one standard language or different languages?

4.4.2.4 Upper-level Ontologies

It can be useful if the ontology library system is 'grounded' in any existing upper-level ontologies, such as Upper Cyc Ontology, SENSUS, MikroKosmos, the PENNMAN Upper Model, and IEEE upper-layer ontology. The upper-level ontology captures and models the basic concepts and knowledge that could be re-used in creating new ontologies and in organizing ontology libraries.

4.4.3 Current Storage Systems

We have surveyed a number of existing ontology library systems, to analyse the current state-of-the art in ontology library systems. The systems that we included in our survey are: WebOnto,[12] Ontolingua,[13] DAML Ontology library system, SHOE,[14] Ontology Server from Vrije Universiteit, Brussels, Belgium,[15] IEEE Standard Upper Ontology,[16] OntoServer[17] and ONIONS.[18] Not all are real ontology library systems, but each provides at least some aspects of a library system. There are other ontology library systems than those that we included in our comparison. We have only included approaches that are publicly available as those offer enough detailed information to enable us to evaluate their actual functionality. The results frequently have been related to Sesame to identify the extensions that might be required in the future. The following summarizes the features of the above-mentioned ontology library systems.

4.4.3.1 Management

Storage. The ontology library systems in this survey fall into one of two

---

[9] http://www.w3.org/rdf/
[10] http://www.w3.org/xml/
[11] http://www.ontoknowledge.org/oil/oilhome.shtml
[12] http://eldora.open.ac.uk:3000/webonto
[13] http://www-ksl-svc.stanford.edu:5915/
[14] http://www.cs.umd.edu/projects/plus/shoe/
[15] http://www.starlab.vub.ac.be/research/dogma/ontologyserver.htm
[16] http://suo.ieee.org/refs.html
[17] http://ontoserver.aifb.uni-karlsruhe.de/
[18] http://saussure.irmkant.rm.cnr.it/onto/

categories: (a) those with a client/server-based *architecture* aimed at enabling remote accessing and collaborative editing (WebOnto, Ontolingua, DAML Ontology Library); and (b) those that feature a web-accessible architecture (SHOE, IEEE SUO). Ontology Server features a database-structured architecture. Most ontologies in this survey are classified or indexed. They are stored in a modular structured library (or lattice of ontologies). WebOnto, Ontolingua and ONIONS all highlight the importance of a modular structure in an ontology library system as that structure facilitates the task of re-organizing ontology library systems and re-using and managing ontologies.

Identification. The standard way to identify an ontology is by its Unique name or Identifier.

Versioning. Only SHOE supports versioning for handling the dynamic changes of ontologies. Versioning is an important aspect of the ontology library system. Although many of the systems surveyed do not currently have this function, they clearly show that it is needed for future improvements.

Sesame. It has a client/server-based architecture and supports web access. It does not support collaborative editing. It deploys the RAL that makes it possible to implement Sesame on top of a wide variety of database-structured repositories without changing any of Sesame's other components. The ontologies stored in Sesame are not classified and are not in a modular structure. Sesame supports ontology versioning, as described in Chapter 11.

4.4.3.2 Adaptation

Searching. Most of these ontology library systems can be accessed through the Internet or World Wide Web. They offer simple browsing only. Ontolingua is the only one that offers some functional searching features, such as keyword searching (wild-card searching), simple query answering, context sensitive searching, etc. As it is embedded in the database management system, Ontology Server could also provide SQL-based searching.

Editing. Most ontology library systems only provide simple editing functions. WebOnto and Ontolingua support collaborative ontology editing (asynchronous and synchronous).

Reasoning. Very simple reasoning functions are provided by WebOnto (rule-based reasoning), Ontolingua (ontology testing) and SHOE (ontology revision).

Sesame. It supports the highly expressive querying of RDFS and RDF data enabled by RQL. It has some simple editing functions but not collaborative editing. The reasoning function of Sesame is described in Chapter 11.

4.4.3.3 Standardization

Language. These ontology library systems use different languages to store their ontologies. In this case, the important function for the future ontology library system should support inter-language translating (like Ontolingua) or some standard language should be accepted or proposed within the ontology community (such as DAML+OIL).

Upper-level Ontology. Ontolingua has a public version of CYC upper-level ontology called HPKB-UPPER-LEVEL with some modification drawn from Pangloss, WordNet, and Penma. WebOnto and SHOE do not have the standard upper-level ontology but have their own fine-grained structure (e.g. Base Ontology). IEEE SUO tries to set up a public standard upper-level ontology.

Sesame. It supports RDF, RDFS and DAML+OIL. It does not have any upper-level ontology.

4.4.4 Requirements for a Storage System

In this section, we summarize important requirements for structuring an ontology library system to enhance ontology management, adaptation and standardization. In doing this, we formulate a wish-list for an ideal ontology library system.

4.4.4.1 Management

Storage. A client/server-based *architecture* is critical to an ontology library system's capacity to support collaborative ontology editing. An ontology library system should also be web accessible.

It is necessary to *classify* an ontology in an ontology library system in order to facilitate searching, managing and re-using ontology. Some of the ontology classification mechanisms available are based on distinguishable features of ontologies. Examples include the following:

- the *subject* of ontologies: the DAML ontology library system classifies ontologies according to the Open Directory Category (www.dmoz.org);
- the *structure* of the ontology: the Ontolingua ontology library system has an inclusion lattice showing the inclusion relations between different ontologies;
- inter and intra ontology *features*: Visser and Bench-Capon (1998) indexed ontologies based on the intra and inter ontology features. Examples include general, design process, taxonomy, axioms, inference mechanism, application, contributions, etc.;

- the *lattice* structure: Noy and Hafner (1997) built a lattice of ontologies showing the relevance of ontologies;
- the *dimensions* of the ontology: Heijst et al. (1997) indexed ontologies using dimensions (task/method dependency and domain dependency) to partition the library into a core library and a peripheral library;
- *stratified upper-level* ontology: ONIONS used generic, intermediate and domain layer to index ontologies;
- the *relations* of ontology: Visser and Bench-Capon (1998) indexed ontology based on defined relations, such as the subset/superset relation, extension relation, restriction, and mapping relation;
- the *components* of ontology: Visser and Bench-Capon (1998) also mentioned the indexing of ontology based on components of ontologies, such as domain partitioning (partition domain in logical units), alternative domain views (polymorphic refinement), abstraction (abstract and detailed ontologies), primary ontologies versus secondary ontologies, terminological, information and knowledge modelling ontologies.

Support for *modular organization* in the ontology library system allows complex ontologies to be composed from simpler ones. This requires maximal cohesion within modules and minimal interaction between modules (McGuinness et al., 2000). Most of the ontology library systems that aim to facilitate ontology re-use, ontology mapping and integration have adopted this structure. ONIONS also highlights the *stratified* design of an ontology library system. Different *naming policies* assist the ontology library system to achieve the modular organization or stratified storage of ontologies (McGuinness et al., 2000). The disjointed partitioning of classes can facilitate modularity, assembling, integrating and consistency checking of ontologies. If, for instance, a certain class, such as 'people,' were disjointed from another class, say 'countries', then consistency checks could be carried out much sooner and faster. Thus, the partition modification has proven to be extremely valuable for editing purposes. Linking class names with their own contexts or using namespaces for differentiating them can serve to prevent violation within individual ontologies. As ontologies continue to grow, so too does the importance of systematic and consistent naming and organizational rules.

Identification. Unique ontology URL, identifier and name are used as the identifier for ontologies in the ontology library systems.

Versioning. A version control mechanism is very important to an ontology library system. Unfortunately, most existing ontology library systems cannot support it, except for SHOE.

4.4.4.2 Adaptation

Searching and editing. An ontology library system should feature a visualized browsing environment, using hyperlinks or cross-references to closely related information. It should support collaborative editing and offer advanced searching features by adopting various existing information retrieval techniques, database searching features, or AI heuristic techniques. Ontology library systems could also monitor user profiles based on access patterns in order to personalize the view of ontologies (Domingue and Motta, 1999).

Reasoning. A simple reasoning function should be included in order to facilitate ontology creation, ontology mapping and integration.

4.4.4.3 Standardization

Language. Syntactically, an ontology representation language should be standardized or inter- or intra-ontology language translation should be supported. Semantically, an ontology library system should feature the *common vocabulary* (or faceted taxonomy). At any rate, it should eliminate the implicitness and misunderstanding of terms in different ontologies (due to synonyms, homonyms, etc.) for most generic classes. Preferably, an ontology library system should also support compatibility with or mapping between multiple controlled vocabularies from different domains. This would not only serve to guarantee flexibility in expressing an ontology semantically, but also to liquidate implicitness. The structures of these common vocabularies or multiple controlled vocabularies must be faceted, or modulated so as to facilitate the reuse, mapping and integration of ontologies (McGuinness, 2000). These vocabularies can help in simple synonym matching, sibling analysis, and disjoint partition checking.

Upper-level ontology. A standard upper-level ontology is important for better organization of ontology library systems (Ontolingua, IEEE SUO).

4.4.4.4 Other requirements

Ontology scalability. Ontology library systems should also consider increasing the scale of ontologies.

Maintenance facility. Ontology library systems should also provide some maintenance features, such as consistency checking, diagnostic testing, support for changes, and adaptation of ontologies for different applications.

Explicit documentation. Each ontology in an ontology library system should be extensively documented. The documentation should include such information as how the ontology was constructed, how to make extensions and what the ontology's naming policy, organizational principles and functions are. Such explicit documents about the ontologies themselves will pave the way for efficient ontology management and re-use.

4.5 Summary

In this chapter, we have looked at various aspects of ontology management. Ontology management is the whole set of methods, methodologies, and techniques that is necessary to efficiently use multiple variants of ontologies from possibly different sources for different tasks.

Alignment is an important aspect, because in many real-world scenarios, there are several ontologies of a domain used for a specific task. Each of those domain ontologies might capture specific aspects of knowledge and might use different terminology. Special mapping ontologies must be created to link different terminologies and modelling styles used in these domain specific ontologies. We described a meta-ontology that can be used to creating such bridges between separated pieces of knowledge. These bridges along with domain ontologies can then be used to perform cross-ontology tasks.

We also discussed the fact that ontologies are not static, but evolve over time. Domain changes, adaptations to different tasks, or changes in the conceptualization require modifications of the ontology. The evolution of ontologies causes interoperability problems which might hamper their effective re-use. Ontology comparison techniques can help the ontology engineer to find changes between ontologies and to characterize them conceptually. When the conceptual relation between the versions, the transformations between them, as well as the meta-data of the change is maintained, it is possible to support both loss-less transformations between version *and* re-interpretation of data and knowledge under different versions.

Ontology library systems are systems that support the ontology management task in various aspects. We have discussed the functions of a ontology library system, we have surveyed exiting systems and finally came up with a wish-list for the ideal ontology library system.

5

Sesame: A Generic Architecture for Storing and Querying RDF and RDF Schema

Jeen Broekstra, Arjohn Kampman and Frank van Harmelen

The resource description framework (RDF) (Lassila and Swick, 1999) is a W3C recommendation for the formulation of meta-data on the World Wide Web. RDF Schema (RDFS) (Brickley and Guha, 2000) extends this standard with the means to specify domain vocabulary and object structures. These techniques will enable the enrichment of the Web with machine-processable semantics, thus giving rise to what has been dubbed the Semantic Web.

We have developed Sesame, an architecture for storage and querying of RDF and RDFS information. Sesame allows persistent storage of RDF data and schema information, and provides access methods to that information through export and querying modules. It features ways of caching information and offers support for concurrency control.

This chapter is organized as follows: In Section 5.2 we discuss why a query language specifically tailored to RDF and RDFS is needed, over and above existing query languages such as XQuery. In Section 5.3 we look at Sesame's modular architecture in some detail. In Section 5.4 we give an overview of the SAIL API and a brief comparison to other RDF API approaches. Section 5.5 discusses our experiences with Sesame to date, and Section 5.6 looks into possible future developments. Finally, we provide our conclusions in Section 5.7.

5.1 The Need for an RDFS Query Language

RDF documents and RDF schemata can be considered at three different levels of abstraction:

- at the *syntactic level* they are XML documents;[1]
- at the *structure level* they consist of a set of triples;
- at the *semantic level* they constitute one or more graphs with partially predefined semantics.

We can query these documents at each of these three levels. We briefly consider the pros and cons of doing so for each level in the next sections. This leads us to conclude that RDF and RDFS should really be queried at the semantic level. We also briefly discuss RQL, a language for querying RDF(S) documents at the semantic level, which has been implemented in the Sesame architecture.

5.1.1 Querying at the Syntactic Level

Any RDF model (and therefore any RDF schema) can be written down in XML notation. It would therefore seem reasonable to assume that we can query RDF using an XML query language (e.g., XQuery) (Chamberlin et al., 2001).

However, this approach disregards the fact that RDF is not just an XML notation, but has its own data model that is very different from the XML tree structure. Relationships in the RDF data model that are not apparent from the XML tree structure become very hard to query.

As an example, the XML description of the RDF model in Figure 5.1 is as follows:

```
<rdf:Description rdf:about=
"http://www.enersearch.se/
corporate/organisation/hohome.html">
  <s:hasName>Hans Ottosson</s:hasName>
  <s:worksFor rdf:resource =
  "http://www.enersearch.com"/>
</rdf:Description>

<rdf:Description rdf:about=
"http://www.enersearch.com">
  <s:legalName>Enersearch AB</s:legalName>
```

---

[1] Actually, this is not necessarily true; non-XML syntaxes for RDF exist, but XML is the most widely used syntax for RDF.

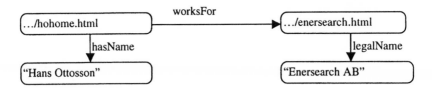

Figure 5.1 An example RDF graph

```
<rdf:type rdf:resource=
"http://www.description.org/
schema#Consortium"/>
</rdf:Description>
```

In an XML query language such as XQuery (Chamberlin et al., 2001), expressions to traverse the data structure are tailored towards traversing a node-labelled tree. However, the RDF data model is a graph, not a tree, and moreover, both its edges (properties) and its nodes (subjects/objects) are labelled. In querying at the syntax level, this is literally left as an exercise for the query builder: one cannot query the relation between the resource signifying 'Hans Ottoson' and the resource signifying 'Enersearch AB' without knowledge of the syntax that was used to encode the RDF data in XML. Ideally, we would want to formulate a query like 'Give me all the relationships that exist between Hans Ottosson and Enersearch AB'. However, using only the XML syntax, we are stuck with formulating an awkward query like 'Give me all the elements nested in a Description element with an about attribute with value "http://www.enersearch.se/corporate/organisation/hohome.html", of which the value of its resource attribute occurs elsewhere as the about attribute value of a Description element that has a nested element title with the value "Enersearch AB".'

Not only is this approach inconvenient, it also disregards the fact that the XML syntax for RDF is not unique: the same RDF graph can be serialized in XML in a variety of ways. This means that one query will never be guaranteed to retrieve all the answers from an RDF model.

5.1.2 Querying at the Structure Level

When we abstract from the syntax, any RDF document represents a set of triples, each triple representing a statement of the form Subject-Predicate-Object. A number of query languages have been proposed and implemented that regard RDF documents as such a set of triples, and that allow such a triple set to be queried in various ways.

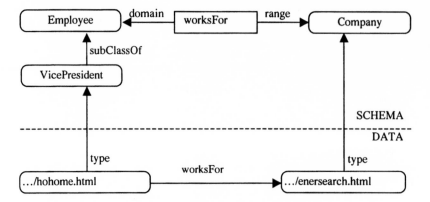

Figure 5.2   An example RDFS graph

Consider the RDFS graph in Figure 5.2. An RDF query language such as, for example, Squish (Miller, 2001) would allow us to query which resources are known to be of type `VicePresident`:

```
SELECT ?x
FROM    somesource
WHERE  (rdf::type ?x VicePresident)
```

The clear advantage of such a query is that it directly addresses the RDF data model, and it is therefore independent of the specific syntax that has been chosen to represent the data.

However, a disadvantage of any query language at this level is that it interprets *any* RDF model only as a set of triples, including those elements that have been given a special semantics in RDFS. For example, since .../hohome.html is of type `VicePresident`, and since `VicePresident` is a subclass of `Employee`, .../hohome.html is also of type `Employee`, by virtue of the intended RDFS semantics of `type` and `subClassOf`. However, there is no triple that explicitly asserts this fact. As a result, the query:

```
SELECT ?x
FROM    somesource
WHERE  (rdf::type ?x Employee)
```

will fail because the query only looks for explicit triples in the store, whereas the triple (../hohome.html, type, Employee) is not explicitly present in the store, but is implied by the semantics of RDFS.

Notice that simply expanding the query into something like:

```
SELECT ?x
FROM    somesource
WHERE   (rdf::type ?x ?c1),
        (rdfs::subClassOf ?c1 Employee)
OR      ?c1 ~ Employee
```

will solve the problem in this specific example, but does not solve the problem in general.

5.1.3 Querying at the Semantic Level

What is clearly required is the means to query at the *semantic* level, that is, querying the full knowledge that a RDFS description entails and not just the explicitly asserted statements. There are at least two options to achieve this goal:

- Compute and store the closure of the given graph as a basis for querying.
- Let a query processor infer new statements as needed per query.

While the choice of an RDF query language is, in principle, independent of the choice made in this respect, the fact remains that most RDF query languages have been designed to query a simple triple base, and have no specific functionality or semantics to discriminate between schema and data information.

RQL (Alexaki et al., 2000; Karvounarakis et al., 2000) is a proposal for a declarative query language that does explicitly capture these semantics in the language design itself. The language has been initially developed by the Institute of Computer Science at FORTH,[2] in Heraklion, Greece, in the context of the European IST project MESMUSES.[3] We briefly describe the language here; for a detailed description of the language see Karvounarakis et al. (2000) and Broekstra and Kampman (2001a).

RQL adopts the syntax of OQL (Cattel et al., 2000), and like OQL, the language is defined by means of a set of core queries, a set of basic filters, and a way to build new queries through functional composition and iterators.

The core queries are the basic building blocks of RQL, which give access to the RDFS specific contents of an RDF triple store. RQL allows queries such as Class (retrieving all classes), Property (retrieving all properties) or Employee (returning all instances of the class with name Employee). This last query of course also returns all instances of subclasses of Employee, since these are also instances of the class Employee, by virtue of the semantics of RDFS. Notice that in RQL, these semantics are defined in the query language itself: the formal query language definition makes a

[2] See http://www.ics.forth.gr/
[3] See http://cweb.inria.fr/projects/mesmuses/

commitment to interpret the semantics of RDFS. This is notably different from an approach like Squish, where the designer/implementer is at liberty to interpret the RDFS entailment using one of the options mentioned earlier, or not at all.

RQL can also query the structure of the subclass hierarchy. In our example, the query subClassOf(Employee) would return the class VicePresident as its only result. In general, this would return all direct and indirect subclasses of Employee, since RQL is aware of the transitivity of the subclass relation. The query subClassOf^(Employee) would return only the immediate subclasses.

For composing more complex queries, RQL has a select-from-where construction. In the from-clause of such a query, we can specify a *path expression*. These allow us to match patterns along entire paths in RDF/RDFS graphs. For example, the query:

```
select Y, $Y
from   VicePresident{X}.worksFor{Y : $Y}
```

returns all things that employ a vice-president, as well as the type of that thing, effectively doing pattern-matching along a path in the graph of Figure 6.2. Notice that RQL path expressions explicitly enable free mixing of data and schema information.

5.2 Sesame Architecture

An overview of Sesame's architecture is shown in Figure 5.3. In this section we give a brief overview of the main components.

For persistent storage of RDF data, Sesame needs a scalable repository. Naturally, a data base management system (DBMS) comes to mind, as these have been used for decades for storing large quantities of data. In these decades, a large number of DBMSs have been developed, each having their own strengths and weaknesses, targeted platforms, and APIs. Also, for each of these DBMSs, the RDF data can be stored in numerous ways.

As we would like to keep Sesame DBMS-independent and it is impossible to know which way of storing the data is best fitted for which DBMS or which application domain, all DBMS-specific code is concentrated in a single architectural layer of Sesame: the *storage and inference layer* (SAIL).

This SAIL is an application programming interface (API) that offers RDF-specific methods to its clients and translates these methods to calls to its specific DBMS. An important advantage of the introduction of such a separate layer is that it makes it possible to implement Sesame on top of a wide variety

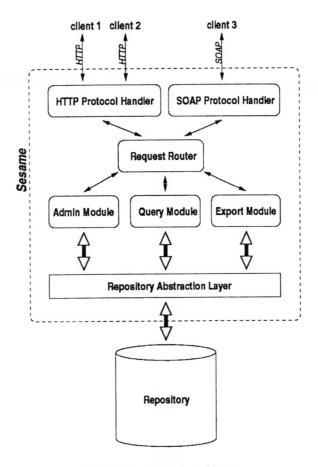

Figure 5.3   Sesame's architecture

of repositories without changing any of Sesame's other components. Section 5.4 looks at the API in more detail.

Sesame's functional modules are clients of the SAIL API. Currently, there are three such modules: the *RQL query engine*, the *RDF admin module* and the *RDF export module*. Each of these modules is described in more detail in the next three sections.

Depending on the environment in which it is deployed, different ways to communicate with the Sesame modules may be desirable. For example, communication over HTTP may be preferable in a Web context, but in other contexts protocols such as remote method invocation (RMI) or the simple object access protocol (SOAP) (Box et al., 2000) may be more suited.

In order to allow maximal flexibility, the actual handling of these protocols has been placed outside the scope of the functional modules. Instead, protocol handlers are provided as intermediaries between the modules and their clients, each handling a specific protocol. The introduction of the SAIL and the protocol handlers makes Sesame into a generic architecture for RDFS storage and querying, rather than just a particular implementation of such a system. Adding additional protocol handlers makes it easy to connect Sesame to different operating environments. The construction of concrete SAIL implementations is discussed in Section 5.4.

Sesame's architecture has been designed with extensibility and adaptability in mind. The possibility of using other kinds of repositories has been mentioned before. Adding additional modules or protocol handlers is also possible.

5.2.1 The RQL Query module

As we have seen, one of the three modules currently implemented in Sesame is an RQL query engine. In Sesame, a version of RQL was implemented that is slightly different from the language proposed by Karvounarakis et al. (2000). The Sesame version of RQL features better compliance to W3C specifications, including support for optional domain and range restrictions as well as multiple domain and range restrictions. It does not, however, feature support for data typing as proposed in the original language proposal. See Broekstra and Kampman (2001b) for details.

The Query Module follows the path depicted in Figure 5.4 when handling a query. After parsing the query and building a query tree model for it, this model is fed to the query optimizer, which transforms the query model into an equivalent model that will evaluate more efficiently. These optimizations mainly consist of a set of heuristics for query sub-clause move-around. Notice that these pre-evaluation optimizations are not dependent on either the domain or storage method.

The optimized model of the query is subsequently evaluated in a streaming fashion, following the tree structure into which the query has been broken down. Each object represents a basic unit in the original query and evaluates

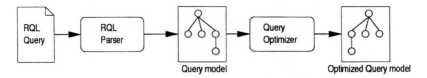

Figure 5.4   Query parsing and optimization model

itself, fetching data from the SAIL where needed. The main advantage of this approach is that results can be returned in a streaming fashion, instead of having to build up the entire result set in memory first.

In Sesame, RQL queries are translated (via the object model) into a set of calls to the SAIL. This approach means that the main bulk of the actual evaluation of the RQL query is done in the RQL query engine itself. For example, when a query contains a semi-join operation over two sub-queries, each of the sub-queries is evaluated, and the semi-join operation is then executed by the query engine on the results.

Another approach would be to directly translate as much of the RQL query as possible to a query specific for the underlying repository. An advantage of this approach is that, when using a DBMS, we would get all its sophisticated query evaluation and optimization mechanisms for free. However, a large disadvantage is that the implementation of the query engine is directly dependent on the repository being used, and the architecture would lose the ability to easily switch between repositories.

This design decision is one of the major differences between Sesame and the RDF Suite implementation of RQL by ICS-FORTH (see Alexaki et al., 2000). The RDF Suite implementation relies on the underlying DBMS for query optimization. However, this dependency means that RDF Suite cannot as easily be transported to run on top of another storage engine.

A natural consequence of our choice to evaluate queries in the SAIL is that we need to devise several optimization techniques in the engine and the SAIL API implementation, since we cannot rely on any given DBMS to do this for us.

5.2.2 The Admin Module

In order to be able to insert RDF data and schema information into a repository, Sesame provides an admin module. The current implementation is rather simple and offers two main functions:

- incrementally adding RDF data/schema information;
- clearing a repository.

Partial delete (on a per statement basis) functionality is not yet available in the current admin module, but support for this feature is under development.

The admin module retrieves its information from an RDF(S) source (usually an online RDF(S) document in XML-serialized form) and parses it using a streaming RDF parser (currently, we use the ARP RDF parser that is part of the Jena toolkit (Carrol and McBride, 2001). The parser delivers the information to the admin module on a per statement basis: (Subject, Predicate, Object). The admin subsequently tries to assert this statement into the repository by communicating with the SAIL and reports back any errors or warnings that might have occurred.

The current implementation makes no explicit use of the transaction functionality of SAIL yet, but we expect to implement this in the near future.

5.2.3 The RDF Export Module

The RDF Export Module is a very simple module. This module is able to export the contents of a repository formatted in XML-serialized RDF. The idea behind this module is that it supplies a basis for using Sesame in combination with other RDF tools, as all RDF tools will at least be able to read this format.

Some tools, like ontology editors, only need the schema part of the data. On the other hand, tools that do not support RDFS semantics will probably only need the non-schema part of the data. For these reasons, the RDF Export Module is able to selectively export the schema, the data, or both.

5.3 The SAIL API

The SAIL API is a set of Java interfaces that has been specifically designed for storage and retrieval of RDFS-based information. The main design principles of SAIL are that the API should:

- Define a basic interface for storing RDF and RDFS in, and retrieving and deleting RDF and RDFS from (persistent) repositories.
- Abstract from the actual storage mechanism; it should be applicable to, for example, RDBMSs, file systems, or in-memory storage.
- Be usable on low-end hardware like PDAs, but also offer enough freedom for optimizations to handle huge amounts of data efficiently on, for example, enterprise level database clusters.
- Be extendable to other RDF-based languages like DAML+OIL (Horrocks et al., 2001).

Other proposals for RDF APIs are currently under development. The most prominent of these are the Jena toolkit (Carrol and McBride, 2001) and the Redland Application Framework (Beckett, 2001). SAIL shares many characteristics with both approaches.

An important difference between these two proposals and SAIL is that the SAIL API specifically deals with RDFS on the retrieval side: it offers methods for querying class and property subsumption, and domain and range restrictions. In contrast, both Jena and Redland focus exclusively on the RDF triple set, leaving interpretation of these triples as an exercise to the user. In SAIL, these RDFS inference tasks are handled internally. The main reason for this is that there is a strong relationship between the efficiency of the inference and the actual storage model being used. Since any particular SAIL implementation has a complete understanding of the storage model (e.g. the database schema in the case of an RDBMS), this knowledge can be exploited to infer, for example, class subsumption more efficiently.

Another difference between SAIL and other RDF APIs is that SAIL is considerably more lightweight: only four basic interfaces are pre-defined, offering basic storage and retrieval functionality and transaction support, but not much beyond that. We feel that in some applications such minimalism may be preferable to an API that has more features, but is also more complex to understand and implement.

The current Sesame system offers several implementations of the SAIL API. The most important of these is the SQL92SAIL, which is a generic implementation for SQL92 (ISO, 1999). The aim is to be able to connect to any RDBMS while having to re-implement as little as possible. In the SQL92SAIL, only the definitions of the data types (which are not part of the SQL92 standard) have to be changed when switching to a different database platform. The SQL92SAIL features an inference module for RDFS, based on the RDFS entailment rules as specified in the RDF Model Theory (Hayes, 2001). This inference module computes the schema closure of the RDFS being uploaded, and asserts these implicates of the schema as derived statements. For example, whenever a statement of the form (`foo`, `rdfs:domain`, `bar`) is encountered, the inference module asserts that (`foo`, `rdf:type`, `property`) is an implied statement.

The SQL92SAIL has been tested in use with several DBMSs, including PostgreSQL[4] and MySQL[5] (see also Section 5.5).

An important feature of the SAIL (or indeed of any API) is that it is possible to put one on top of the other. The SAIL at the top can perform some action when the modules make calls to it, and then forward these calls to the SAIL beneath it. This process continues until one of the SAILs finally handles the actual retrieval request, propagating the result back up again.

We implemented a SAIL that caches all schema data in a dedicated data structure in main memory. This schema data is often very limited in size and is requested very frequently. At the same time, the schema data is the most difficult to query from a DBMS because of the transitivity of the subClassOf and subPropertyOf properties. This schema-caching SAIL can be placed on

---

[4] See http://www.postgresql.org/
[5] See http://www.mysql.com/

top of arbitrary other SAILs, handling all calls concerning schema data. The rest of the calls are forwarded to the underlying SAIL.

Another important task that can be handled by a SAIL is concurrency handling. Since any given RQL query is broken down into several operations on the SAIL level, it is important to preserve repository consistency over multiple operations. We implemented a SAIL that selectively blocks and releases read and write access to repositories, on a first come first served basis. This set-up allows us to support concurrency control for any type of repository.

5.4 Experiences

Our implementation of Sesame can be found at http://sesame.aidministrator.nl/, and is freely available for non-commercial use. The implementation follows the generic architecture described in this chapter, using the following concrete implementation choices for the modules:

- We use both PostgreSQL and MySQL as database platforms. The reason we are running two platforms simultaneously is mainly a development choice: we wish to compare real-life performance of both platforms and SAIL implementations.
- We have various repository set-ups running, combining different stacks of SAILs (including the SQL92SAIL, the PostgreSQL SAIL, the MySQL SAIL, and a schema cache and a concurrency handler) on top of each repository.
- A protocol handler is realized using HTTP.
- The admin module uses the ARP RDF parser.

In this section, we briefly report on our experiences with various aspects of this implementation.

5.4.1 Application: On-To-Knowledge

Figure 5.5 shows how Sesame serves as the central data repository for a number of tools in our knowledge management architecture as outlined in Chapter 1:

- OntoExtract extracts ontological conceptual structures from natural language documents. OntoWrapper performs the same task for semi-structured information sources. These ontologies are uploaded for storage in Sesame.
- The resulting ontologies can be downloaded into the OntoEdit ontology

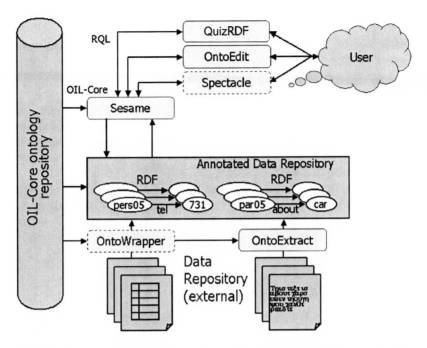

Figure 5.5 Sesame is positioned as a central tool in the knowledge management architecture

editor. When the user has edited an ontology, the result is again stored in Sesame.

- The resulting ontologies are downloaded into QuizRDF, a semantic search engine, which provides search, browse and query facilities for web data based on the ontologies.
- Spectacle generates web sites whose content and navigation structure are based on the ontologies and data stored in Sesame.
- OntoShare allows end-users in a community to share knowledge using the ontological structure stored in Sesame. An RDF-annotated information resource for the community is built up automatically and stored in Sesame.

Because Sesame is a server-based application, the integration of all this functionality is realized simply by establishing HTTP connections to Sesame. We are currently in the process of applying this architecture in a number of knowledge management applications.

5.4.2 RDFS in Practice

While developing Sesame, many ambiguities in the RDFS specification were uncovered. One of the reasons for this is that RDFS is defined in natural language: no formal description of its semantics is given. As a result of this, the RDFS specification even contains some inconsistencies.

In an attempt to solve these ambiguities, the RDF Core Working Group has been chartered to revise the RDF and RDFS specifications. One of the results is a formal Model Theory for RDF (Hayes, 2001), which specifies model and schema semantics more precisely and includes a formal procedure for computing the closure of a schema.

As mentioned in Section 5.4, the SQL92SAIL features an inference module that follows the procedure described in the model theory. Our experiences are that a naive implementation of this formal procedure is painfully slow, but with relative ease it can be optimized to perform quite satisfactorily. Improving this performance even further is currently work in progress.

5.4.3 PostgreSQL and SAIL

In our first test set-up for Sesame we used PostgreSQL. PostgreSQL is a freely available (open source) object-relational DBMS that supports many features that normally can only be found in commercial DBMS implementations.

One of the main reasons for choosing PostgreSQL is that it is an object-relational DBMS, meaning that it supports sub-table relations between its tables. As these sub-table relations are also transitive, we used these to model the class and property subsumption relations of RDFS.

The SAIL that is used in this set-up therefore is specifically tailored towards PostgreSQL's support for sub-tables (which is not a standard SQL feature). It uses a dynamic database schema that was inspired by the schema shown in Karvounarakis et al. (2000). New tables are added to the database whenever a new class or property is added to the repository. If a class is a sub-class of other classes, the table created for it will also be a sub-table of the tables for the super-classes. Likewise for properties being sub-properties of other properties. Instances of classes and properties are inserted as values into the appropriate tables. Figure 5.6 gives an impression of the contents of a database containing the data from Figure 5.2.

The actual schema involves one more table called `resources`. This table contains all resources and literal values, mapped to a unique ID. These IDs are used in the tables shown in the figure to refer to the resources and literal values. The `resources` table is used to minimize the size of the database. It ensures that resources and literal values, which can be quite

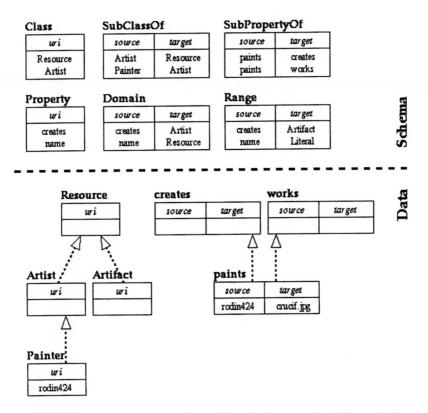

Figure 5.6   Impression of the object-relational schema used with PostgreSQL

long, only occur once in the database, saving potentially large amounts of memory.

In the test set-up, several optimizations in the SAIL implementation were made, such as selective caching of namespaces and frequently requested resources to avoid repetitive table lookups.

Our experiences with this database schema on PostgreSQL were not completely satisfactory. Data insertion is not as fast as we would like. Especially incremental uploads of schema data can be very slow, since table creation is very expensive in PostgreSQL. Even worse, when adding a new subClassOf relation between two existing classes, the complete class hierarchy starting from the subclass needs to be broken down and rebuilt again because sub-table relations cannot be added to an existing table; the sub-table relations have to be specified when a table is created. Once created, the sub-table relations are fixed. Another disadvantage of the sub-table approach is that cycles in the class hierarchy cannot be modelled properly in this fashion.

In a new set-up, we used the SQL92SAIL to connect to PostgreSQL. The current version of this SAIL implementation takes a radically different approach: all RDF statements are inserted into a single table with three columns: Subject, Predicate, Object. While we have yet to perform structured testing and analysis with this approach, it seems to perform significantly better, especially in scenarios where the RDFS changes often.

For querying purposes, the original PostgreSQL SAIL performed quite satisfactorily, especially when combined with a schema-caching SAIL stacked on top (see Section 5.4). We have yet to perform structured testing on querying with the new SQL92SAIL, but initial results show that it performs somewhat slower than the earlier PostgreSQL SAIL, which is to be expected.

5.4.4 MySQL

In initial tests with MySQL, we implemented a SAIL with a strictly relational database schema (see Figure 5.7). As can be seen, a number of dependencies arise due to the storage of schema information in separate tables. In order to keep overhead to a minimum, every resource and literal is encoding using an integer value (the id field), to enable faster lookups. To encode whether a

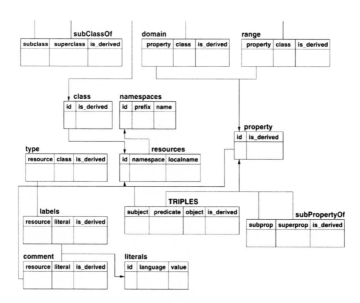

Figure 5.7   Impression of the relational schema used with MySQL

particular statement was explicitly asserted or derived from the schema information, an extra column is_derived is added where appropriate.

The main difference between this schema and the schema used in the PosgreSQL set-up (see Figure 5.6) is that in this set-up, the database schema does not change when the RDFS changes. In application scenarios where the RDFS (the ontology) is unstable, this is an advantage because typically adding new tables to a database requires more time and resources than simply inserting a row in an existing table.

As in the PostgreSQL SAIL, selective caching of namespaces and other optimization techniques were implemented in this set-up. Overall, this approach performed significantly better in our test scenarios, especially on uploading.

5.5 Future Work

5.5.1 Transaction Rollback Support

While the SAIL API has support for transactions, it currently has no transaction rollback feature. Transaction rollbacks, especially in the case of uploading information, are crucial if we wish to guarantee database consistency. In the case of RDF uploads, transaction rollbacks can be supported at two levels:

- A single upload of a set of RDF statements can be seen as a single transaction, or alternatively, a single upload can be 'chunked' into smaller sets to support partial rollback when an error occurs during the upload session.
- A single RDFS statement assertion can be seen as a transaction in which several tables in the database need to be updated. From the user point of view, the schema assertion is atomic ('A is a class'), but from the repository point of view, it may consist of several table updates, for instance, in the schema presented in Figure 5.6, a new table would have to be created, and new rows would have to be inserted into the 'Resources' and the 'Classes' table.

Both levels of transaction rollback support may help ensure database consistency. Together with the concurrency support already present in the Sesame system, this will help move Sesame towards becoming an ACID[6] compliant storage system (note, however, that this can only be guaranteed if the platform used for storage supports it).

---

[6] Atomicity, concurrency, isolation, durability – these four properties of a transaction ensure database robustness over aborted or (partially) failed transactions.

5.5.2 Versioning Support

Chapter 11 describes an implementation of per statement versioning. This basic type of versioning will enable more elaborate versioning schemes.

5.5.3 Adding and Extending Functional Modules

Sesame currently features three functional modules. We plan to extend the functionality of these modules, as well as add new modules. In the current admin module implementation, only incremental upload of RDF statements is supported. We plan to implement more advanced update support, most importantly support for deleting individual triples from the repository. A prototype implementation of this new feature already exists but has to be tested and extended further. Plans for new modules include a graphical visualization component and query engines for different query languages (for example, Squish).

5.5.4 DAML+OIL Support

As mentioned in Section 5.4, the RDF SAIL API has been designed to allow extension of the functionality, for example, to include support for DAML+OIL. Such an implementation is described in Chapter 11.

5.6 Conclusions

In this chapter we have presented Sesame, a generic architecture for storing and querying both RDF and RDFS information. Sesame is an important step beyond the currently available storage and query devices for RDF, since it is the first publicly available implementation of a query language that is aware of the RDFS semantics.

An important feature of the Sesame architecture is its abstraction from the details of any particular repository used for the actual storage. This makes it possible to port Sesame to a large variety of different repositories, including relational databases, RDF triple stores, and even remote storage services on the Web.

Sesame itself is a server-based application, and can therefore be used as a remote service for storing and querying data on the Semantic Web. As with the storage layer, Sesame abstracts from any particular communication protocol, so that Sesame can easily be connected to different clients by writing different protocol handlers.

We have constructed several concrete implementations of the generic architecture, using PostgreSQL and MySQL as repositories and using HTTP as communication protocol handlers.

Important next steps to expand Sesame towards a full fledged storage and querying service for the Semantic Web include implementing transaction rollback support, versioning, extension from RDFS to DAML+OIL and implementations for different repositories. This last feature especially will be greatly facilitated by the fact that the current SAIL implementation is a generic SQL92 implementation, rather than specific for a particular DBMS.

6

Generating Ontologies for the Semantic Web: OntoBuilder

R.H.P. Engels and T.Ch. Lech

6.1 Introduction

Significant progress has been made in technologies for publishing and distributing knowledge and information on the web. However, much of the published information is not organized, and it is hard to find answers to questions that require more than a keyword search. In general, one can say that the web is organizing itself. Information is often published in relatively ad hoc fashion. Typically, concern about the presentation of content has been limited to purely layout issues. This, combined with the fact that the representation language used on the World Wide Web (HTML) is mainly format-oriented, makes publishing on the WWW easy, giving it an enormous expressiveness.

People add private, educational or organizational content to the web that is of an immensely diverse nature. Content on the web is growing closer to a real universal *knowledge base*, with one problem relatively undefined; the problem of the interpretation of its contents. Although widely acknowledged for its general and universal advantages, the increasing popularity of the web also shows us some major drawbacks. The developments of the information content on the web during the last year alone, clearly indicates the need for some changes. Perhaps one of the most significant problems with the web as a distributed information system is the difficulty of finding and comparing information.

Thus, there is a clear need for the web to become more *semantic*. The aim of introducing semantics into the web is to enhance the *precision* of search, but also enable the use of *logical reasoning* on web contents in order to answer queries. The CORPORUM OntoBuilder toolset is developed specifically for this task. It consists of a set of applications that can fulfil a variety of tasks, either as *stand-alone* tools, or augmenting each other.

Important tasks that are dealt with by CORPORUM are related to document and information *retrieval* (find relevant documents, or support the user finding them), as well as information *extraction* (building a knowledge base from web documents to answer queries), information dissemination (summarizing strategies and information visualization), and automated document classification strategies.

First versions of the toolset are encouraging in that they show large potential as a supportive technology for building up the *Semantic Web*. In this chapter, methods for transforming the current web into a *semantic web* are discussed, as well as a technical solution that can perform this task: the CORPORUM tool set. First, the toolset is introduced; followed by some pragmatic issues relating to the approach; then there will be a short overview of the theory in relation to CognIT's vision; and finally, a discussion on some of the applications that arose from the project.

6.1.1 OntoBuilder and its Relation to the CORPORUM System

When generating and building up, utilizing and maintaining a semantic web, the main raison d'être for such a web is the fact that people need to deal with information overload. Naturally, this holds at the individual level when learning, organizing and interacting on the Internet as much as for organizational learning, knowledge sharing and socializing (Bremdal et al., 1999).

Facilitation of these scenarios by the CORPORUM system requires a diverse functionality, covering a broad variety of applications. Scenarios differ from information collection, categorization, analysis and extraction to browsing, visualization and navigation aids. In order to give such a semantic web some autonomy, intelligent agents can be defined for automation of many tasks. CORPORUM serves as the 'brains' of such intelligent agents, gathering intelligence of all kinds on the Internet, intranets or networked servers, in order to find specific information. This information could be medical knowledge for a specific new medicine, a student wanting to collect material for a course or business intelligence about potential market opportunities or threats. For such intelligent agent scenarios, web server components, database servers, mission schedulers and a client server component are included in CORPORUM (Bremdal and Johansen, 2000). The CORPORUM tool set comprises a component that is able to generate textual summaries based on information distribution in discourses, visualize graphical summaries of texts and relate personal profiles to summary generation (Bremdal, 2000). By offering a server for semantic analysis of natural language (free text) texts it effectively puts the flesh on the bones of the Semantic Web (Bremdal and Johansen, 2000; Engels and Bremdal, 2000). Analyses of structured and unstructured texts are performed by OntoBuilder. The basic principles on which the linguistic approach is based are explained in Section 6.3.

In short, three main scenarios for applying the CORPORUM system can be identified:

1. *extraction* of information from texts for building knowledge bases (cf. semantic web);
2. *retrieval* of information from other sources (search scenarios); and
3. strategies to *compress*, *visualize* and *disseminate* information to people (dissemination and navigation).

It is at the extraction layer that OntoBuilder is defined. It consists of the two modules OntoExtract and OntoWrapper.

The OntoBuilder toolbox will eventually be able to extract all kinds of information from structured sources (OntoWrapper) and unstructured free text (OntoExtract). These documents could be available on your local intranet or the World Wide Web.

6.1.2 OntoExtract

OntoExtract builds on a core natural language analysis engine. The engine supports analysis of natural language texts and generates lightweight, domain specific ontologies of these texts while utilizing already existing knowledge from a central data repository.

OntoExtract utilizes highly expressive querying of RDF data and schema information, using RQL, which is being developed by the ICS-FORTH institute in Greece. Sesame (described in the previous chapter), is used for the repository management. The Sesame data repository can be questioned through the access layer and edited with the OntoEdit ontology editor described in Chapter 7.

The overall extraction process, shown in Figure 6.1, is typically iterative, with a user initiating the process by providing a location to analyse for seed-ontology generation, together with some parameters that define the extent of the Internet domain to be analysed. From each page containing natural language text, lightweight ontologies are produced and automatically submitted to the Sesame repository for storage. The statements that are

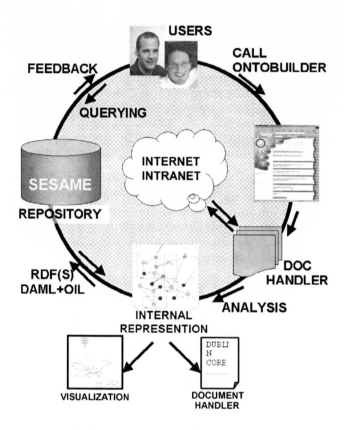

Figure 6.1   The OntoBuilder extraction cycle

submitted include pointers to the original documents from which the knowledge is extracted. Figure 6.2 provides an example of generated output from the
case study described in Chapter 13.

The toolset can be used in two basic scenarios. In one scenario, the user has
no starting point for ontology building and is actively pursuing one. OntoExtract can now analyse a few documents that are seen as 'core' with respect to
the domain that is to be modelled. The user provides the location(s) of the
domain after which the software analyses it. After the analysis phase an initial
ontology is uploaded to the Sesame server. OntoEdit can be used from that
point in order to visualize, alter and extend this initial ontology according to its
needs. When analysing documents, OntoExtract keeps track of which
concepts are already available in Sesame, and will output concepts closely
related to those already stored and confirmed. The user can request these

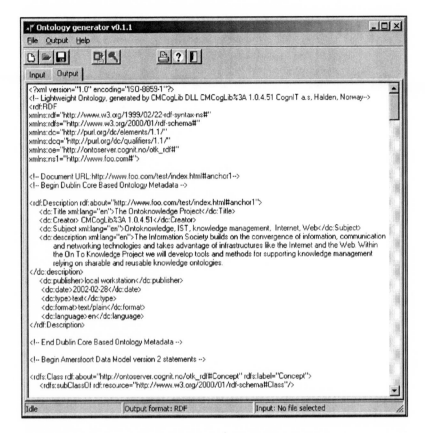

Figure 6.2   Excerpt of output generated by OntoExtract on the Enersearch domain

concepts to be visualized as a 'proposal' for ontology extension within OntoEdit.

Secondly, there is the scenario in which the ontology builder has sufficient overview of the domain that is to be modelled. In that case, the starting point will be a user-defined initial ontology, which is inputted using OntoEdit and stored in Sesame. OntoExtract can now retrieve this ontology from the Sesame repository and take this into consideration as 'background knowledge' when analysing more documents. In this scenario, OntoExtract will only propose concepts that are semantically 'close' to the already existing ontology, either being closely related to existing concepts, or being sub-/super-classes, extension definitions of classes and so on.

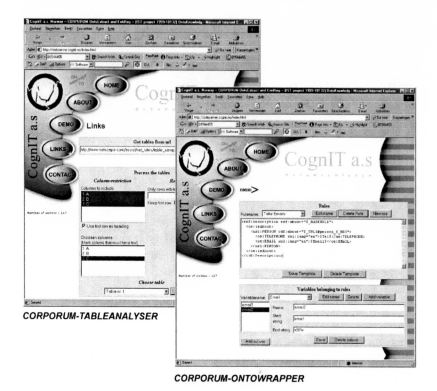

CORPORUM-TABLEANALYSER

CORPORUM-ONTOWRAPPER

Figure 6.3 OntoWrapper and TableAnalyser allow for the analysis of structured web pages and table based content extraction

6.1.3 OntoWrapper and TableAnalyser

Often, an ontology engineer would like to analyse (web) domains for instance data in order to populate ontologies. Instance data is frequently represented as structured information. This means that in many scenarios, there is a need for a tool that can be configured for such tasks. OntoWrapper and TableAnalyser are defined for these tasks (see Figure 6.3).

OntoWrapper deals with the analysis of structured pages and allows for the definition of rule-based entity extraction. OntoWrapper allows the user to define XML/RDF templates, variables and rule sets in order to perform a structured analysis of a specific domain (e.g. a company's internal phone directory). It iteratively scrolls through all the available pages and extracts, for example, the names of people and their telephone numbers. The information that is found is exported to the RDF repository.

Figure 6.4   Template defining an instance of class 'person' used by OntoWrapper

The ontology engineer defines RDF templates and OntoWrapper then analyses the specified documents, automatically generating the merged output and sending it to the Sesame repository as data statements about specific pages.

Figure 6.4 provides an example RDF template, to be automatically filled out with information extracted from a phone directory. OntoWrapper explicitly populates the repository with data according to the ontology that is defined and stored in Sesame. In the current example, the class ⟨CLASS: PERSON⟩ is defined to have three properties ⟨PROPERTY: NAME⟩, ⟨PROPERTY: TELEPHONE⟩ and ⟨PROPERTY: EMAIL⟩. For every analysed instance that is found in the web domain, the system automatically generates a new instance of class 'PERSON' which is automatically numbered.

OntoBuild is dependent on a 'document handler' which recursively crawls the Internet, thereby avoiding loops, reiterations and other issues related to automated web-crawling. The current component can deal with a variety of formats, ranging from HTML, ASCII and PDF to documents in MS-Office format.

6.2 Reading the Web

6.2.1 Semantics on the Internet

The disadvantage of 'flat', format-based representation languages such as HTML, is that they rarely combine information about content (the text a writer wants to disseminate), and layout (the format in which this is to be done). In the case of HTML, this disadvantage has to be set against a rather easy to learn language, so that virtually anybody with web access can easily publish information, knowledge and opinions. Using an explicit representation language with clear semantics, usually puts an end to ease of use for most average users. In addition, the web is a medium for publishing content and much of this content is not originally created with web-based publishing in mind. This, once more, shows the need for the introduction of a clearly separated semantic representation of content.

History teaches that few scenarios evolve so completely as to totally obliterate their past. An example of this would be the development of web browsers over the years, where backward compatibility has always been a major issue. Seen from such a point of view, there are two different scenarios in which the World Wide Web could evolve further. On the one hand, content and semantics could be extracted from the currently existing mass of documents available on the web, thus probably reproducing the initial semantics of the web. Alternatively, there is possibility of starting by defining the *semantics* in a page and developing web sites from there on.

The former alternative implies a need for redundant storage, as web contents are stored in their original form (e.g. HTML annotated) as well as in a *semantic* form (e.g. in the form of RDF code that can be stored inside a page or in a separate repository like Sesame (see Chapter 5).

The latter alternative will in many cases imply a breach with the 'old style' and probably imply the need for a new and different architecture for the World Wide Web. As is often the case, each of these approaches has its own drawbacks.

Given that a combination of approaches is often the best way to go, one can imagine tool support in order to either analyse pages that are not represented in a 'semantically rich' manner, or graphical interfaces (editors) to support creating such semantic representations (semi-) automatically (cf. Chapter 7). A variety of projects developing and evaluating semantic representation languages have shown that representing all web content in 'higher order languages' might not be feasible unless more automated approaches become available. Several projects have been initiated worldwide to support this Semantic Web approach on all levels (Fensel et al., 1999; Hendler, 2000). This includes defining languages, extensions on languages and query languages (Brickley and Guha, 2000; Fensel et al., 2000) and the implementation of tools for (semi-) automatic content extraction (Bremdal and Johansen, 2000; Knowledge Management Group, 2000).

6.2.2 Problems with Retrieving Natural Language Texts from Documents

Most approaches for natural language analysis start at a level where 'clean' text is available. Often there are also restrictions on the scope of such 'clean' text in that it may only contain a single discourse. Besides problems related to the linguistic analysis of a piece of text, the actual physical retrieval process can be rather problematic. Regularly, it is seen that theoretically fine approaches do not make it in the light of real-world applications, simply because the pre-processing steps are not implemented satisfactorily. Thus, understanding the often pragmatic problems during retrieval, as well as defining ways to overcome them, can lift accuracy in the practical application of such linguistic technologies dramatically.

What type of problems can be identified? Generally speaking, the following factors strongly influence the results of applied linguistic systems:

- *Document handling*: although a rather trivial issue, we mention it for completeness sake. The factors that follow are all strongly dependent on the ability to get a running text from a document. Not all document formats allow for the extraction of such running texts (as streams of characters and layout information), or at least make them very hard to get.
- *Normalization*: documents can be parsed up front during retrieval such that special characters, line breaks, paragraph breaks, hyphenation and the like are represented in a standard manner; bulleted lists and enumerations are normalized; and such that as few as possible differences between the final analysed texts exist.
- *Multiple discourses*: one of the harder problems is to find out how many discourses a document actually contains. Is it a single, consistent document or is it a mere collection of smaller, independent pieces of information? As soon as linguistic approaches start to perform co-reference chaining, pronoun resolution and the like, there is a clear need for a thorough discourse boundary analysis.
- *Document categorization*: closely related to the issue of multiple discourses is the issue of document classification. Ideally, one is able to analyse up front the type of document that is going to be retrieved. In that way, it is possible to treat it accordingly and possibly sub-divide it.
- *Writing style*: perhaps one of the least emphasized facts in linguistic analysis on the Internet is the fact that the vast majority of authors are non-native speakers. This has a profound effect on analysis methods that are mostly based upon the assumption that texts are grammatically correct.
- *Layout issues*: texts can be found in various locations within a document, e.g. in headings, captions, and within pictures. Font size, type and appearance can convey meaning (bold, italics, etc.).

These six factors relating to the early stages of document analysis strongly influence results in the later stages. Some effects are easily dealt with, others are really domain-specific and some accumulate problems. We now take a closer look at these factors.

6.2.3 Document Handling

There are an increasing number of documents that can only be read with third-party programs and plug-ins. These are documents that do not limit themselves to the standard HTML format and its derivatives (XHTML, DHTML, etc.), and to standard scripting languages (ASP, JavaScript, CGI, etc.). Due to the sheer variety of programs in this area, and the fact that each community and usage seems to have its own preferred programs, standardization is a seemingly fruitless effort. Problems with retrieving documents from the Internet are multifold.

Linguistic analysis really benefits from information on important words, sentences and paragraphs. One can think of titles, abstracts, emphasized words and phrases and so on. Every document format has its own way of representing such information, and whereas some formats allow easy identification of this information, others do not.

Many programs and plug-ins do provide ways to export the original document into ASCII or another less complex format that is easily usable. Layout information is often lost, which is one thing, but often the text is also obstructed. Exporting a two-column text from some document formats can mean that the exported ASCII represents the document as if it was a one-column text. Similar problems are often found with images, captions under images that appear in the middle of a text, page numbers and other items on the page that are completely mixed with the text. It does not need arguing that severe problems will arise in the later phases of the document analysis.

6.2.4 Normalization

Related to these non-standardized representation issues is the issue of 'text normalization'. As many linguistic algorithms take standard (enhanced) ASCII as input, this input has to be generated out of the various formats, possibly using a variety of (proprietary) components and plug-ins. Normalization is necessary to generate the pre-specified format required by the particular parser being used. Normalization needs to take account of all the different ways of representing, for example, headlines, sections, line breaks, but also paragraphs, chapters, etc. It also needs to take account of the various ways and characters used to achieve hyphenation, for example, for word hyphenation or for the identification of compounds, etc. Finally, normalization is required to overcome the differences in standards used in the document format, for example, the identification of a bulleted list in HTML vs. PDF.

6.2.5 Multiple Discourses

Discourse analysis can be done on several levels. A natural division is between an analysis on the document level (is a document built up of smaller documents or not?), or at the linguistic level (how many discourses does the current text contain?). In this section we are concerned with the document level, that is, how much information is available for deciding whether a currently retrieved document is about a single or multiple topics. Documents like newsletters, news front pages, listings, but also contracts and legal documents often discuss a variety of topics in one single document, whereas research papers, articles and so on often basically discuss a single domain or topic.

Analysis of discourses at this level will obviously be very layout oriented, with heuristics used for recognition of articles in news pages. A specific order of events and characters can be enough to identify a headline:

$$<CR>+<LF>+<EMPHASIZED\ TEXT>+<CR>+(OPTIONAL:$$
$$<LF>)\ +<SMALLER\backslash\ FONT>\Rightarrow HEADLINE$$

Other document information that can lead to the identification of specific text parts are HTML anchor and SPAN tags, used for identification of text within a document as in:

$$<A\ NAME="\#987457">HEADLINE</A>$$

Similar heuristics can be implemented for documents in MS-Office format, where in many cases paragraphs and sentences already are classified according to a set of tags with properties of their own (e.g. heading styles, body text styles, etc.). Many such identifiers can be used in discourse identification at the document level, but there are again only a few heuristics that are universally applicable to all pages.

Probably the best way to tackle this sheer diversity of ways to identify discourses is to combine this step with the later step of discourse analysis at the linguistic level. In that case, it is advantageous to retain as much format and layout information as possible, to be available at later stages of analysis.

6.2.6 Document Class Categorization

Related to the previous issue is the problem of document classification. It may be possible to classify documents at the document level, so that general parsing strategies with class-specific heuristics can be applied.

This is probably most clearly seen when looking at news pages vs. insurance documents or contracts. Whereas the former document type usually distinguishes between articles through visual clues in the form of emphasized text, bulleted lists or headlines, contracts and insurance documents often identify a breach in discourse by an enumeration in its articles (i.e. '§1.2', '§1.3', etc.). One common approach is to use the 'Naive Bayes' classifier, which involves a little basic probability and very little guidance from a human director.

6.2.7 Writing Style

Specific styles of writing by different authors are an important factor in explaining results from an automated analysis process. The way in which a piece of prose is written determines for a large part how 'interpretable' a text is. Many of the factors that play a role in this are really cognitive. People use several strategies to explain themselves, from discourses that explain things bottom up, top down, from the beginning or just from some intermediate 'knowledge' level or even not at all. Additionally one can try to explain a specific phenomenon straight away, or by using an analogy. Texts are not only written to explain but also to influence (like political statements, adverts, etc.). Some texts are also written for a particular group of readers (children, specialists, youth, etc.). All these factors determine how much additional background knowledge is needed in order to read and understand a discourse. It is expected that if documents can be classified up front (see Section 6.2.6), the kind of analysis and preprocessing and the amount of knowledge needed to really understand the text can be estimated. Common strategies for dealing with these topics are unfortunately not known, whereas a variety of approaches tackle partial problems in specific domains. It is to be expected that until these issues are resolved, many approaches will have to fall back on more general linguistically oriented heuristics.

6.2.8 Layout Issues

Currently, many approaches first completely normalize texts, thereby throwing away much of the non-textual conventions on layout, which contains much information of importance and relevance. A typical analysis process might benefit from knowledge about headlines, the relations between footnotes and text markers, the paragraph a specific caption belongs to, ingresses and their relation to the whole document, and also font sizes, emphasized texts and so on. Often there are pieces of text that could be filtered out (e.g. program snippets in a text on programming), or should be treated differently (e.g. citations). There are approaches that aim at tackling specific issues in layout, whereas the overall topic of layout interpretation is probably closely related to document classification.

It is probably infeasible to expect heuristics based on layout to apply in general, the more since this issue is one of cognition and culture rather than being purely technical.

6.3 Information Extraction

Information extraction (IE) is the task of obtaining structured information from unstructured sources, for example, natural language text (Klabunde et al., 2001). IE-related tasks are by no means restricted to text, as there are numerous domains like tables, lists, etc. However, the following section focuses on natural language text only. With the Internet developing rapidly and huge amounts of information being available, the necessity for systems that organize information in a more accessible manner is obvious. Thus, IE-related applications are found in a variety of fields (Engels and Bremdal, 2000):

- Abstracting and summarizing: aims to deliver shorter, informative representations of larger documents.
- Visualization of text: documents can often be visualized according to the concepts and relationships that play a role.
- Comparison and search (information retrieval): find instances of semantically similar pieces of information.
- Question answering: in computer-human interacting systems, such as expert systems.
- Induction/deduction of knowledge based on extracted information: many approaches from the field of machine learning play a role here.
- Knowledge base generation: information that is extracted, deduced or induced can be used in other scenarios as well. A knowledge base can be regarded as a typical container for transfer or sharing of such knowledge across applications and time.

In this section, we describe the main strategies and linguistic techniques in IE, followed by presentation of the main features and strategies employed by the CORPORUM OntoBuilder toolkit.

6.3.1 Content-driven Versus Goal-driven

Today's IE landscape is mainly characterized by two major strategic approaches. On the one hand, there are systems using goal-driven (top down) techniques for acquiring the information desired and, on the other hand, there are initiatives extracting the information in a content-driven (bottom up) manner.

Goal-driven systems, for example, SMES (Declerck and Neumann, 2000), are generally domain specific and make use of knowledge about the domain in question. The information of interest typically is predefined as open slots, in so-called templates, which are to be filled by the relevant parts of the text. Goal-driven systems often make use of shallow parsing techniques, and this has led to a renaissance of finite state techniques, for example, augmented transition networks or finite state transducers.

Content-driven systems, on the other hand, are designed to work more or less domain-independently. Typically, content-driven systems build up ontologies/taxonomies based on heuristics as well as statistical techniques.

Goal-driven and content-driven systems have in common that both of them are highly dependent on a thorough low-level linguistic analysis of a given text, comprising normalization, tokenization and part-of-speech (POS) tagging.

As of today, many goal-driven systems perform reasonably well within their domains (for evaluation methods, see, e.g. the MUC conferences). However, their dependency on predefined domains is a clear drawback and makes them rather unattractive in the Semantic Web context, or for agents or other domain-independent applications. This drawback has led to initiatives that are either trying to make goal-driven systems more domain-adaptive (e.g. SMES; Declerck and Neumann, 2000), or combining them with content-driven techniques (e.g. GETESS; Düsterhöft et al., 1999).

On the other hand, the domain-independence of content-driven systems comes with a deficiency in terms of performance. The lack of accuracy makes content-driven systems somewhat inappropriate as tools for extracting information units that require a high degree of precision (e.g. knowledge base generation). However, their fuzziness is an important feature when it comes to tasks dealing with uncategorized texts such as summarization or agent-based searching.

6.3.2 Levels of Linguistic Analysis

Like other tasks within the field of natural language processing (NLP), the performance of IE systems is dependent on a linguistic analysis of the texts in question. However, there is a high degree of variation among different systems in terms of the analysis level and thoroughness. While some systems try to perform an exhaustive analysis over all linguistic levels, others rely only on a rough lexical analysis. In the following, we give a short survey of the different levels and their relevance in the IE context. Of course, this survey cannot give more than a shallow impression of the complexity of the challenges at each level. Since today's IE applications mainly work with documents and other written text, speech recognition and other tasks on the phonological level are omitted in this context.

6.3.2.1 Tokenization

Tokenization is the task of dividing a text into single lexical tokens. It is an absolutely vital task for any IE system that goes beyond very primitive forms of crude pattern matching, and a task that is often underestimated in its complexity. Besides simple white space identification, tokenization includes activities such as dehyphenation, sentence boundary detection, proper name recognition, and – for multi-lingual systems – language detection. There is a range of different techniques for solving these problems. Whereas some systems employ statistical techniques (Palmer and Hearst, 1994) for disambiguating sentence boundaries (the 'full stop vs. abbreviation period' problem), others only make use of heuristics like regular expressions (Grefenstette and Tapanainen, 1994).

6.3.2.2 Lexical/Morphological Analysis

After tokenization, most IE systems perform a lexical analysis on the tokenized text. The main task at this level is the POS tagging. POS tagging is a prerequisite for dividing grammatical words like determiners (the, a, those) or prepositions, from lexical words like nouns, verbs or adjectives, since the latter tend to be the content-bearing units in a text. Stemming of inflected words is a major challenge in this context, raising the problem of deciding between a full form lexicon or tools for morphological analysis. Most of today's POS tagging systems use statistical methods as well as heuristics for this task, employing both suffix-stripping algorithms, word form tables and frequency lists. In addition to the pure POS tagging, many IE systems perform other forms of morphological analysis on the tokens, for example, compounds analysis.

6.3.2.3 Syntactic Analysis

The goal of syntactic analysis is to determine the underlying structure in a string of words. There is a wide range of grammar formalisms applied in computational linguistics. On the one hand, there are declarative unification-based formalisms like, for example, lexical functional grammar (LFG) (Kaplan and Bresnan, 1982) or head-driven phrase structure grammar (HPSG)

(Pollard and Sag, 1994). On the other hand, there are procedural methods like finite state parsers, or augmented transition networks. After having been criticized for their shortcomings in linguistically correct description of language, the latter have experienced a renaissance recently, since many goal-driven IE systems employ augmented transition networks (Klabunde et al., 2001) and other shallow parsing techniques.

Furthermore, there are systems employing a combination of the techniques mentioned above, using statistics and finite state techniques for chunk parsing, and advanced grammar formalisms on a whole sentence level. However, context independent parsing of full sentences is computationally highly expensive, especially when being applied to less formal texts, and therefore not yet implemented in large-scale commercial IE systems.

6.3.2.4 Semantic/Pragmatic Analysis

A typical semantic analysis of a sentence will return the sentence's proposition(s) in a formal language, e.g. predicate logic. Most of the advanced grammar formalisms supply their own tools for describing a sentence's meaning (e.g. the Σ-structure in LFG), or use so called Φ roles (theta roles, or semantic roles) to describe the meaning of the constituents in a sentence.

Other approaches to semantics have their root in the frame-based work of Winograd or Schank, for example, memory based parsing (Lebowitz, 1983). These systems utilize knowledge about prototypical situations and the involved participants. Many of those systems use script mapping (scenario mapping) in order to identify the participants in given instances of the situations.

6.3.2.5 Discourse Analysis

Discourse analysis is a vast and still mainly unconquered territory in linguistics. Thus, there are not many systems that work with a large-scale implementation of computational discourse analysis. There are many challenges within the field of discourse analysis, like, for example, the problem of discourse boundaries. Especially for IE purposes, it is essential to be able to distinguish between two different instances of a given event type. Other tasks related to discourse analysis include the detection of digressions, enumerations, introduction of new topics and entities. Another challenge within this field is pronoun resolution and co-reference chaining.

6.3.3 CognIT Vision

CognIT's technology supports both personal and enterprise wide document and information management – that is, management by content. The CORPORUM system is based on CognIT's core technology developed in Norwegian research labs. This technology focuses on meaningful content: CognIT's mission is to capture content with respect to the interest of the individual rather than address the document itself. There are three essential aspects of this:

- CORPORUM interprets text in the sense that it builds ontologies that reflect world concepts as the user of the system sees and expresses them. The ontology constitutes a model of a person's interest or concern. The interest model is applied as a knowledge base in order to determine contextual and thematic correspondence with documents presented before it.
- The interest model and the text interpretation process drive an information extraction process that characterizes the hit in terms of relevance and in terms of content. This information can be stored in a persistent database for future reference.

CORPORUM is able to augment 'meaning' structures with concepts that are invented from the text. The core component of the CORPORUM system is also able to extract the information most pertinent to a specific text for summary creation, and extract the so-called core concept area (CCA) from a text. Results are represented according to a ranking which is based on interest relative to a specific contextual theme set by the user. On top of that, the CORPORUM system is able to generate explanations, which will allow the user to make an informed guess on which documents to look at and which to ignore. CORPORUM can point to exactly those parts of the targeted documents that are most pertinent to a specific user's interest.

In order to accomplish these tasks, CORPORUM OntoExtract (part of CORPORUM OntoBuilder) employs techniques at different levels of linguistic analysis. Like other IE systems CORPORUM performs a thorough lexical analysis including text normalization, tokenization, and POS tagging.

CORPORUM performs no extensive syntactic parsing, at any rate not within the framework of any of the major grammar formalisms. Skipping the syntactic processing at whole sentence level makes CORPORUM fast and robust as it does not require a priori selection of a particular IE technique.

At the semantic level, CORPORUM performs word sense disambiguation, by describing the context in which a particular word is used. This is naturally closely related to knowledge representation issues. In addition, CORPORUM extracts content bearing words and collocations, weighting them according to a scoring algorithm. These concepts and the weighted connections between them are the basis for the lightweight ontologies created by the OntoBuilder.

6.4 Knowledge Generation from Natural Language Documents

Natural language processing and information extraction are fields that have reached a specific level of maturity, which is starting to show in the increasing presence of commercial and freely available technology. Currently, speech recognition and generation is used in telephone interaction with human beings; OCR software is used for recognition of hand-written documents; tools for spelling correction are available; and much more is about to appear on the market. The CORPORUM toolset extracts and generates semantic structures from documents. These are then exported in the form of graph-like images and ontologies. From each encountered document, the system extracts meta-data that can be stored in the Sesame repository.

6.4.1 Syntax Versus Semantics

Over the past 30 years there has been an ongoing discussion on what focus to sustain, syntax or semantics, in order to pursue natural language processing. This discussion has surfaced in both the empiricist camp as well as the rationalist camp. Yet we see a tendency that both are important. There must be significant interplay between syntax and semantic modelling. In fact the whole issue of language understanding circulates in our view around the process of communication. In terms of written expressions this implies that it is not sufficient to focus on the text alone, but on the whole system of author, author's knowledge, author's goals and constraints, author's world, chosen vocabulary, listener's context and listener's knowledge. The work of Ram and Moorman (1999) is a recent and quite impressive contribution in this direction.

6.4.2 Generating Semantic Structures

The description of the issues and processes involved in bringing an analysis of a natural language text from the token level to a semantic level is very basic to the understanding of the level of granularity of the generated semantic structures (Engels and Bremdal, 2000). The outcome of the analysis process is a virtual representation of a text's contents, its semantics (e.g. the concepts playing a role in it, the role type, relations between various roles that are identified, and so on).

Generation of semantic knowledge in information extraction is based upon the results of parsing steps that can be of varying 'analysis depth'. Some approaches build on knowledge about word types only, whereas other approaches go beyond that and require a deep understanding of the sentence structure before being able to generate the semantic representations that are aimed at.

When analysing published IE approaches the whole range can be found, from utilizing statistical word lists (often augmented with frequencies that are learned on document corpora), through hierarchies of concepts that are related to each other by specific relation types (concept–superconcept relations, part–whole relations, etc.), towards even richer semantic representations like the theory of conceptual dependencies (Shank, 1975). Generally speaking, such representational schemes are said to represent a world state, certain beliefs or even factual knowledge that is implicitly present in the analysed documents. There are very diverse opinions on how to represent semantics. Some research groups see semantics as the semantics of some logic representation. Others propose that semantics and context are contained in a more 'fuzzy' way by statistical properties over text. Other representation formalisms are also found, but the division between symbolic approaches (representation through logic) and sub-symbolic (connectionist, fuzzy or statistical approaches) is important. In OntoExtract the initial analysed and annotated text is transformed into an internal representation that makes use of a variety of linguistic analysis steps to come to an initial interpretation of what is written. After an additional disambiguation phase, performing co-reference chaining, resolution of, for example, abbreviations, pronoun resolution, etc., an internal representation of the text is built up. This representation contains the original text, its annotations, but also the resolutions performed on it. Now it is possible to tell what is going on, who is involved, and what type of relations are discussed in the discourse that has been analysed.

Whereas this internal representation is a very rich one, the model used contains a mix of properties and contents that is not suitable for a usage in a clean, formal knowledge representation such as DAML+OIL or RDF(S). Therefore the semantic structures undergo a translation such that a more formal representation emerges, filtering out the pragmatic properties and focusing on a clean representation. Other export filters defined in this internal representation format deal with export of Dublin Core meta-data, XML representations of texts, (semantically ordered) key concept vector lists and semantic networks for visualization.

6.4.3 Generating Ontologies from Textual Resources

A number of issues, many of them generally known from the field of knowledge representation, have to be dealt with at this stage. When representing results of this information analysis phase into formal representation paradigms like DAML+OIL (incorporating elements of F-LOGICS and Description Logics, which come with a clear semantics), there is also a need to have a clear philosophy on how the translation from linguistics into such formalisms can be done properly. As would be clear to the observant reader, these issues are not trivial. Among the most pertinent problems that are encountered, we report:

- the problem of representation level, what knowledge should be represented at the ontology level, what knowledge should be represented at the fact level (cf. the general discussion on ontology modelling as in: what represents an 'instance' and what represents a 'concept');
- the problem of dealing with the inheritance problem (e.g. taxonomies);
- the issue of consistency between extracted ontologies and their truth within specific domains.

Whereas the first issue is a general issue that is and has been discussed for a long time in several research fields, the last two issues have recently become more pertinent with the wider availability of web resources and the introduction of automated services for their analysis and representation. It is not easy to decide and find out what the domain of a subset of web documents is, nor is it easy to decide which and how much of the extracted taxonomies/ontologies are related to the same 'world view'. During the course of the project, it became quite clear that an automated approach to this is not quite feasible based on current state-of-the-art technology alone. It might require some intervention from human beings to decide whether extracted interrelations between concepts, properties and instances actually hold for a specific paragraph, text, domain, part of a domain or possibly universally. This aspect of the current approach taken in the On-To-Knowledge project is presumably more a cognitive issue than a linguistic issue.

Currently ontologies are extracted from single documents taken from the web. From these documents concepts are extracted, but also created. These are then set into relation with each other, augmented with properties and eventually found instances are hooked up to them. On top of this, relationships that are considered to be more or less cross-taxonomic (i.e. not representing pure class/super-class or part/whole relations) are included so that an additional reasoning through such concept clusters is also allowed for. One example of

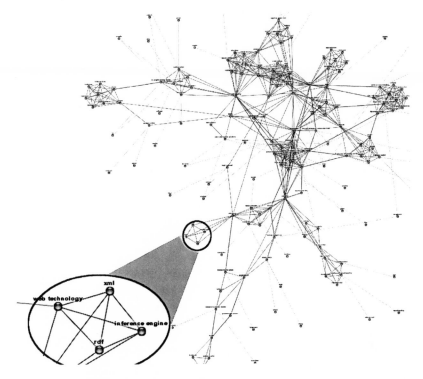

Figure 6.5 CORPORUM generated semantic network structures visualized with CCAviewer[1]

usage of this kind of relationships is that of 'orthogonal' lists of concepts used for providing help with searching in unfamiliar domains.

6.4.4 Visualization and Navigation

Rather different from the more logic-based representations defined in DAML+OIL are the exported semantic network structures. These exported networks, in either an XML or an RDF representation, can be run through a graph layout algorithm in order to generate visualizations (cf. Figure 6.1).

Such visualizations allow for usage in visual browsers and navigators based on larger document sets, and offer people an at-a-glance overview of the information to which they have access. Figure 6.5 shows a simplified structure created from a specific text visualized with CCAviewer.

---

[1] CCAviewer: visualization software module developed by Aidministrator Nederland BV (On-To-Knowledge Consortium Partner).

The enlarged part of the structure shows the semantic cluster around the terms 'XML', 'RDF', 'inference engine'and 'web technology'. There are more of such clusters recognizable, and their intercluster relationships are used to navigate from one cluster to another by relevant concepts. Pictures that are thus automatically generated from natural language texts provide an at-a-glance overview over a piece of information. They can be used in order to augment executive summaries and readers' aids, but they are also used as visual interfaces to databases (preferably in corporate settings). As such they augment knowledge management systems, where they provide a visual entrance to pieces of information pertinent to specific interest groups within an enterprise.

6.5 Issues in Using Automated Text Extraction for Ontology Building using IE on Web Resources

Although many approaches exist that deal with the processing and extraction of information from natural language resources, there is not a final answer to the problem of how to perform that task. Many of the approaches found use statistical methods on larger corpora of documents in order to disclose taxonomic relationships (is-a, part-whole). Such approaches tend to get into trouble when domains are dynamic (and the statistics have to be recalculated often), or whenever no large corpora are present in order to calculate support for rare relations. Providing such relations with a semantically meaningful label is not trivial when using corpora-based, statistical heuristics either. On top of this, the Internet has an additional challenge that is not taken into consideration very often, and that is the multi-cultural background of the authors contributing to it. Many of the more statistically oriented approaches tend to model peculiarities in writing style, which can lead to interesting observations when non-native authors struggle to bring forward a point while using a restricted vocabulary.

A few clear presumptions that were made at the start of this work:

- Semantics on the web should be modelled and represented in an explicit manner.
- Ontologies should be built for a variety of domains, in order to facilitate recall and precision of 'information searches'.
- The approach is not domain-specific.
- Current representation languages are not sufficient, therefore a formal representation language (DAML+OIL) has been developed.
- Many of the current Internet resources should be analysed using automated information extraction techniques that are not too domain specific.

Taking these five main assumptions, considering the literature and reports from academia and taking into account practical experience gained in use of the tools, a variety of observations can be made.

First, transforming 'informal text' into formal representations requires a 'deep' analysis of the text if the aim is to really build formally valid, logically correct knowledge bases. Revealing class/sub-class relationships and part-whole relationships are very difficult. Ways exist to extract such information, but there is a 'social consensus' or 'cognitive' argument that makes it very difficult to really create these pieces of knowledge. An important factor is the 'open world' assumption that holds on the internet.[2]

Related to this problem is the problem of enriching the knowledge that can be generated from a single document resource. Currently, the OntoBuilder software generates knowledge on a per document basis. The system is 'unbiased' towards specific domains, and should therefore be universally applicable, which is one of the more general requirements for web-based NLP technology. Due to this non-specific applicability, the current system has to work on the syntactic/morphological level only, and will encounter difficulties in areas where knowledge about the world is needed.

The current OntoBuilder software analyses on a per page/document basis. Therefore, the system is said to build small, lightweight ontologies for each and every document that is encountered. Within the current project, we have taken the assumption that documents found within a specific domain, are describing the same 'view on the universe'. Therefore all knowledge generated is currently hooked up to the domain from which the documents stem (e.g. the definition of a concept 'car' is hooked up to the domain root of, for example, 'http://www.ford.com'). Specific instances of the concept 'car', such as a report on a specific 'ford mondeo' that crossed the Sahara desert, are found on pages within the domain (e.g. 'http://www.ford.com/travels/sahara/mondeotour.html'). Such specific knowledge is then hooked up through the page it stems from. Obviously, such an approach runs into numerous problems, for which (again), no universal solution exists.

Closely related to this problem of the scope of validity of modelled knowledge, truth and trust are other issues that play an important role. Is a model valid across domains? Can we model a domain in such a way that truth is maintained? Does an aeroplane definition by an engineer in the United Kingdom refer to the same aeroplane like structure as in the definition of an aboriginal in Australia? Or are they two different entities? Is a specific statement, for example, on a war, universally true, or is it only true within the eyes

---

[2] Open World Assumption: no knowledge generated can be falsified, since there is only partial knowledge on the state of the world. For the same reason it is hard to define any class that is universally valid, because not the whole world is known.

of a specific observer? How can we extract the difference out of the documents that are retrieved? The same holds for the treatment of contradictions in descriptions within and between domains, specialization of and extensions to class definitions, similarity and analogy. Basically, these issues all relate to the issue of the scope of discourses and the validity of knowledge across the borders of such a scope.

Then, is it all that bad? Is it maybe not a good idea to use automated natural language technology as Semantic Web technology? Obviously, there are enormous gains to be expected from relatively simple extensions on current technology, such as:

- non-'pure' ontology definition, defining different types of 'cross taxonomic' interrelationships between concepts;
- recognition of simple elements like named entities and their most important relations to other close or nearby concepts;
- automated thesaurus building/analysis of hypernyms/hyponyms to enhance relevant concept recognition.

Automated ontology building without having available extensive world knowledge is probably not feasible when it comes to very rich models, but such generated ontologies can be used as 'seed ontologies', automatically generated from a variety of user defined documents. Seed ontologies can be enriched manually using graphical editors, and stored for later reference. Current extensions of the software incorporate as much knowledge as possible from the repository, recapturing automated extraction in an iterative process, so that an eventual lack of explicit background knowledge can be taken care of.

Within current approaches to the Semantic Web, it is debatable which should be central – the human using the web or the possibility of performing machine processing on web content. In the former case, logic representations are probably not the most intuitive for use with humans, and different, more 'cognitive' representations of such knowledge might be more convenient. In the latter case, there is a clear need for more formal definitions of knowledge representation, but its sheer complexity in relation to some of the problems mentioned above (truth maintenance, contradictions, scope of validity, etc.) might seem a major stumbling block. Another key debate concerns natural language processing technology and which technologies turn out to be the most suitable for the Semantic Web. As long as logic representations (cf. DAML+OIL) are seen as the way to store knowledge in the Semantic Web, it is probably better to aim at formal grammar methods for analysis of rich and logically consistent models. However, it is clear that such approaches tend to suffer from their complexity and richness, in such a way that there will be an enduring need for pragmatic solutions, probably based on statistical, corpora-based technology.

The current OntoBuilder software shows us that there are possibilities to explore, and there is a real chance to get some automated information extraction technology for the Semantic Web in place and the area to look into might be at the interface between the formal approaches and corpora-based approaches. Playing a central role in future research will be challenging issues such as discourse boundary analysis, the scope of the validity of knowledge extracted from such a discourse, document type classification in order to tackle differences in writing and speaking, and others issues related to these.

7

OntoEdit: Collaborative Engineering of Ontologies

York Sure, Michael Erdmann and Rudi Studer

7.1 Introduction

Developing ontologies is central to our vision of Semantic Web-based knowledge management. The methodology described in Chapter 3 guides the development of ontologies for different applications. However, because of the size of ontologies, their complexity, their formal underpinnings and the necessity to come towards a shared understanding within a group of people when defining an ontology, ontology construction is still far from being a well-understood process. In recent years, research has aimed at paving the way for the construction of ontologies by ontology development environments (Duineveld et al., 1999; Fridman et al., 2000; Arprez et al., 2001). Different directions have been taken to support the engineering of ontologies:

1. Several seminal proposals for *guiding* the ontology development process *by engineering methodologies* have been described (Uschold and King, 1995; Lopez et al., 1999), which influenced the ontology development environments (Arprez et al., 2001).
2. *Inferencing mechanisms* for large ontologies have been developed and implemented (e.g. Horrocks, 1998), also to support ontology engineering (Bechhofer et al., 2001).
3. Finally, the need to achieve consensus about an ontology was reflected by *collaborative environments* (Tennison and Shadbolt, 1998; Domingue, 1998; Swartout et al., 1996; Farquhar et al., 1996) for ontology engineering.

However, only a few of these seminal approaches (e.g. Arprez et al., 2001) have worked towards combining all of these urgent desiderata. This observation seems to reflect our own experience, viz. that it is far from trivial to offer a sound integration of these aspects. Therefore, OntoEdit is an ontology engineering environment that is rather unique in its kind as it combines methodology-based ontology development with capabilities for collaboration[1] and a tight integration into a tool framework, i.e. the framework outlined in Chapter 1.

Concerning the methodology, OntoEdit focuses on three of the main steps for ontology development (the methodology is described in Chapter 3), viz. the kick off, refinement, and evaluation. We describe the steps supported by OntoEdit and focus on collaborative aspects that occur during each of the step. First, all requirements of the envisaged ontology are collected during the kick off phase. Typically for ontology engineering, ontology engineers and domain experts are joined in a team that works together on a description of the domain and the goal of the ontology, design guidelines, available knowledge sources (e.g. re-usable ontologies and thesauri, etc.), potential users and use cases and applications supported by the ontology. The output of this phase is a semi-formal description of the ontology. Second, during the refinement phase, the team extends the semi-formal description in several iterations and formalizes it in an appropriate representation language like RDF(S) or, more advanced, DAML+OIL. The output of this phase is a mature ontology (the 'target ontology'). Third, the target ontology needs to be evaluated according to the requirement specifications. Typically this phase serves as a proof for the usefulness of ontologies (and ontology-based applications) and may involve the engineering team as well as end users of the targeted application. The output of this phase is an evaluated ontology, ready for roll-out into a productive environment.

Support for these collaborative development steps within the ontology development methodology is crucial in order to meet the conflicting needs for ease of use and construction of complex ontology structures. We now illustrate OntoEdit's support for each of the supported steps. The examples shown are taken from the Swiss Life case study on skills management (cf. Chapter 12).

7.2 Kick Off Phase

As in software engineering and as proposed by Lopez et al. (1999), we start ontology development by collecting requirements for the envisaged ontology.

---

[1] In a recently submitted paper (Sure et al., 2002), we have described how methodology and inferencing interact to support the ontology engineering process.

By nature, this task is performed by a team of experts in the domain accompanied by experts in modelling. The goal of this phase is to set up requirement specifications for the ontology and to produce a first draft version of the ontology ('semi-formal ontology description').

The outcome of such a requirement specification phase is (i) a document that contains all relevant requirements (domain and goal of the ontology, design guidelines, available knowledge sources, potential users and use cases and applications supported by the ontology), (ii) a semi-formal ontology description, i.e. a graph of named nodes and (un)named, (un)directed edges, both of which may be linked with further descriptive text.

To operationalize our project methodology it is desirable to have tools that reflect and support the steps of the methodology and guides users step by step through the ontology engineering process. Along with the development of the methodology, we therefore extended the core functionalities of OntoEdit with two plug-ins to support the first stages of the ontology development, viz. OntoKick and Mind2Onto.[2] OntoKick targets (i) creation of the requirement specification document and (ii) extraction of relevant structures for the building of the semi-formal ontology description. Mind2Onto targets the integration of brainstorming processes.

As computer science people, we are familiar with software development and prefer to start with a requirement specification of the ontology, that is, OntoKick. People who are not so familiar with software design principles often prefer to start with 'doing something'. Brain storming is a good method to quickly and intuitively start a project, therefore one also might begin the ontology development process with Mind2Onto.

OntoKick supports the collaborative generation of requirements specifications for ontologies. The collaborative aspect of OntoKick is not so much the support for the distributed work of the team members, but rather the support for the personal interaction of the ontology engineers and the domain experts. This is a two step process. First, OntoKick allows important meta-aspects of the ontology to be described, viz. the domain and goal of the ontology, design guidelines, available knowledge sources (e.g. domain experts, re-usable ontologies, etc.), potential users, use cases, and applications supported by the ontology. OntoKick guides the engineering team stepwise through all the relevant aspects and stores these descriptions along with the ontology definitions.

Second, OntoKick supports the creation of a semi-formal ontology description. Naturally, domain experts are a valuable knowledge source for structuring a domain. A very common method for knowledge acquisition from domain

---

[2] Describing the technical foundations of the plug-in framework is beyond the scope of this work; it is described in Handschuh (2001). In a nutshell, one might easily expand OntoEdit's functionalities through plug-ins.

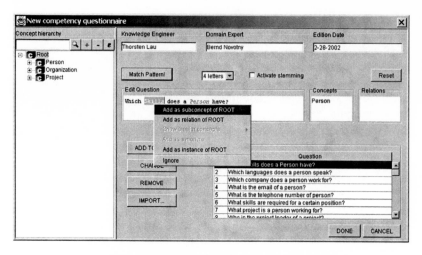

Figure 7.1   OntoKick: capturing of competency questions

experts is personal interviews. To structure the interviews with domain experts, we use competency questions (CQ) (cf. Uschold and King, 1995). Each CQ defines a query that the envisaged ontology (or the ontology-based application) should be able to answer and therefore defines explicit requirements for the ontology. Typically, CQs are derived from interviews with domain experts and help to structure knowledge. We take further advantage by using them to create the initial version of the semi-formal description of an ontology and also for evaluation of the ontology in a later stage. Based on the assumption that each CQ contains valuable information about the domain of the ontology, we extract relevant concepts, relations and instances from these CQs (cf. Figure 7.1).

OntoKick establishes and maintains links between CQs and concepts derived from them (cf. Figure 7.2). This allows for better traceability of the origins of concept definitions in later stages and improves quality assurance during the development process, that is, by documenting the origins and the context of concepts, relations and instances. Therefore, the level of quality of the ontology reached can be re-engineered by other ontology engineers. 'Real life' ontology modelling is supported by OntoKick as the following example from the Swiss Life Skills Management case study illustrates. First, the ontology engineer interviews a domain expert. Thereby they identify CQs, for example, 'Which skills does a person have?' (cf. Figures 7.1 and 7.2). Based on these CQs, the ontology engineer creates a first draft of the semi-formal description of the ontology and models it graphically in OntoEdit. He identifies relevant concepts, relations and instances from the above-mentioned

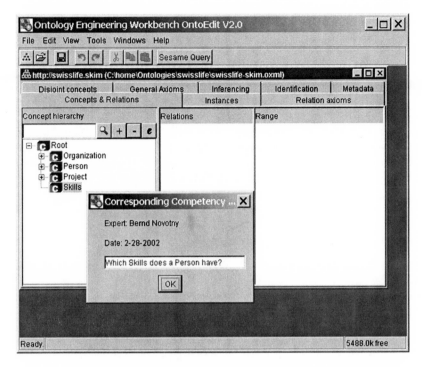

Figure 7.2   OntoKick: traceability of competency questions

CQ, for example, the concept `Person`. The instances might, for example, be prototypical instances that are used for evaluation purposes. After capturing CQs and modelling the ontology with OntoEdit, the ontology engineer is able to retrieve corresponding CQs for each concept, relation and instance. This helps him and others to identify the context in which these elements were modelled.

Mind2Onto is a plug-in for supporting brainstorming and discussion about ontology structures. Especially during the early stages of projects in general, brainstorming methods are commonly used to quickly capture pieces of relevant knowledge. Mind maps are widely used (Buzan, 1974); they were originally developed to support more efficient learning and evolved into a management technique used by numerous companies. Numerous tools already exist for the electronic creation of mind maps. Many people from academia and industry are familiar with mind maps and related tools, including potential ontology engineers and domain experts. Therefore, the integration of electronic mind maps into the ontology development process is very attractive (e.g. Lau and Sure, 2002).

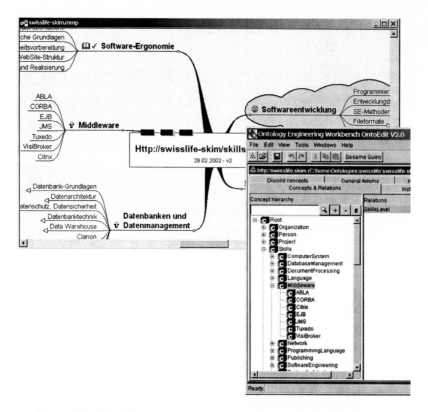

Figure 7.3   Mind2Onto: research topics as a mind map and in the ontology

We have relied on a widely used commercial tool[3] for the creation of mind maps. It supports collaborative engineering of mind maps through peer-to-peer communication and has advanced facilities for graphical presentations of hierarchical structures, for example, easy to use copy and paste functionalities and different highlighting mechanisms. Its strength but also its weakness lies in the intuitive user interface and the simple but effective usability, which allows for quick creation of mind maps but it lacks expressiveness for advanced ontology modelling. By nature, mind maps have no assumptions for its semantics, that is, elements connected by branches are somehow 'associatively related' to each other. This assumption fits perfectly well during the early stages of ontology development for quick and effective capturing of relevant knowledge pieces and makes the mind map tool a valuable add-on.

Mind2Onto integrates the mind map tool into the ontology engineering

---

[3] MindManager 2002 Business Edition; cf. http://www.mindjet.com

methodology. Currently OntoEdit and the mind map tool interoperate through import and export facilities based on XML (cf. Figure 7.3). In the Swiss Life Skills Management case study (cf. Chapter 12) we used the mind map tool to facilitate discussions about the skills hierarchy. Most of the domain experts were already familiar with the tool, the others learned how to use it very quickly. Initially all workgroups created a mind map of relevant skills in joint sessions. The peer-to-peer communication of the mind map tool provided the necessary workgroup functionalities for this effort. Another requirement for the ontology was multi-linguality (viz. German, English, French and Italian). The mind map itself was modelled by German-speaking domain experts who preferred to model in German. We transferred the German concepts and used OntoEdit's facilities to complete the multi-lingual external representations in cooperation with people from the Swiss Life translation service.

7.3 Refinement Phase

The goal of this phase was to refine the semi-formal description of the ontology according to the captured requirements into a mature ontology, which is the output of this phase. We found that, especially during the refinement phase, different teams worked simultaneously on developing the ontologies. For example, several sessions were necessary in the skills management case study until the skill structures were accepted by all members. In this phase, relationships are typically added to refine an ontology (a task that the brainstorming tool is not capable of). After reaching a consensus, the mind map created for skills (see previous section) was restructured in cooperation with the ontology engineers to facilitate the interoperability with OntoEdit; that is, as a simple assumption we took the elements of a mind map as concepts and branches as 'subConceptOf' relationships between concepts (cf. Figure 7.3).

In the current version of OntoEdit, members of an engineering team can collaborate even though they are geographically distributed and still modify the ontology at the same time. We have developed a client/server architecture (cf. Figure 7.4) in which the clients connect to an ontology server and can change or extend the ontology.[4] All clients are immediately informed of modifications performed by the other ontologists. Engineers can store comments (e.g. explaining design decisions) in a documentation field for

---

[4] For practical reasons, our first collaborative version is based on Ontobroker (cf. Decker et al., 1999) as an ontology server (with reasoning capabilities), which was initially developed at the Institute AIFB. The communication with Sesame (cf. Chapter 5) is currently restricted to non-collaborative upload and download of ontologies. For the future, we plan to connect the On-To-Knowledge Ontology Middleware and Reasoning component (cf. Chapter 11) to enable collaborative ontology engineering on top of Sesame.

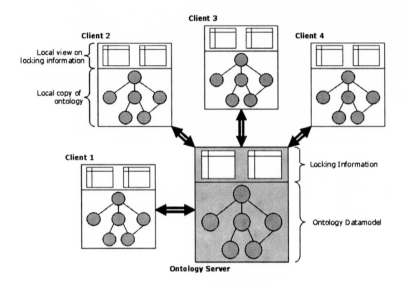

Figure 7.4   Client/server architecture of OntoEdit

each concept and relation. In this way, one of the main features of ontologies, that is, their consensual character, is supported. Collaborating ontologists must agree on the modelling decisions that are made. Therefore it is the possible to monitor the development process of all collaborators, which is essential for reaching the goal of a shared ontology.

7.3.1 Transaction Management

In a distributed development environment, certain mechanisms must be implemented to ensure safe development conditions, such as consistency of the models and the provision of a minimum degree of concurrency. To reach this goal, we employed a locking and transaction protocol and implemented a distributed event model on the basis of Java-RMI (remote method invocation). To guarantee consistent models, the clients are forced to obtain locks for each resource (e.g. concept, instance, relation) that they want to modify (e.g. add a super-concept, add an attribute-value pair to an instance, or change the arity of a relation).[5] The server denies the (write) access to a resource if the resource is not locked by the client that attempts to modify it. Clients can

---

[5] The grounding data model of OntoEdit is OXML 2.0. This frame-based model offers a number of meta-classes, like ontology, concept, relation, but also predicate or axiom, with a rich set of properties and associations to facilitate ontology modelling (cf. http://www.ontoprise.de/download/oxml2.0.pdf for a reference manual).

obtain locks either by explicitly locking these resources, or more conveniently, by a begin of transaction (BOT) that is accompanied with a list of needed resources. If not all resources can be assigned to the calling client, the BOT fails and the transaction is immediately aborted. Otherwise the server locks the needed resources for the client, so that no other client can manipulate them until the end of the transaction is reached.

The client can now manipulate the locked resources until it commits the transaction. After a commit, all locked resources are freed again and the operations performed in the body of the transaction are actually applied to the data model. Afterwards, events are created to inform the other clients of the modifications performed. If the transaction needs to be aborted by the client, all operations are undone, all locks are removed, and no events are fired. Transactions may be nested to make complex operations possible without the need for rollback mechanisms. For example, the data model procedure of moving a concept from one super-concept to another one consists of two subtransactions (remove a super-concept relationship to the first super-concept and establish a new one for the second concept) that must be performed all together or none at all. Because of the necessity for nested transactions, we implemented a strict two phase locking protocol (S2PL).

In this protocol, additional resources can be achieved (and locked) within the body of a transaction. Our implementation of the S2PL allows for arbitrarily nested transactions. The execution of inner transactions and the release of all locked resources is postponed until the outermost commit or abort is finally reached. Again, only after the final commit are events sent to the other clients. We employ the S2PL because (i) it allows for nested transactions and (ii) prevents cascading aborts. Thus, clients can be immediately informed if a planned operation will commit or is prohibited due to unavailable resources. (iii) S2PL also prevents deadlocks since resources are only locked in a BOT if all locks can be achieved. Other locking protocols are either too inflexible (like conservative locking (C2PL) that cannot lock resources in addition to the locks of the BOT and thus, is not suitable for nested transactions) or provide chances for deadlocks that must be appropriately handled.

To reduce communication overhead, save bandwidth and because transactions are relatively short lived, no information about transactions (especially not about locked objects within a BOT) is communicated from the server to other clients; that is, the local view on locking information within a client (cf. Figure 7.4) contains all resources that are locked by this client (by a BOT) but none that have been locked by a BOT of any other client. Nevertheless, another kind of locking information *is*distributed to all clients: An ontologist can lock a whole sub-tree of the concept hierarchy. The server informs all clients of this locking operation.

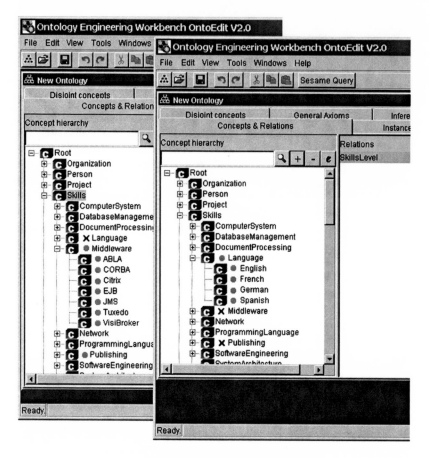

Figure 7.5   Locked trees in OntoEdit

7.3.2 Locking Sub-trees of the Concept Hierarchy

A common practice in ontology engineering is to start with a top level structure and to refine it later on. Different parts of an ontology can be refined by
different ontologists or groups. These collaborators should be able to work on
their parts of the ontology with as little interference from other ontologists as
possible. This is achieved in OntoEdit by the possibility of locking a complete
sub-tree of the concept hierarchy. After the sub-trees have been locked, no
conflicts can arise anymore, and what is equally important, the need to check
for locking information *with the server* is reduced drastically. Since most
modelling operations will occur within the scope of the sub-trees, that is,

will mainly access already locked resources, the client can decide *locally* whether these operations are permitted or not. This (tree) locking information is distributed to all other clients and visually indicated in the GUI (cf. Figure 7.5). Crosses mark concepts that are locked by other clients and may not be edited. Bullets mark concepts that may be edited, altered and removed at will.

Due to the distribution of this information, clients can often check locally whether a transaction will be permitted or not. If all needed resources are marked as 'locked by me' in the local view of the locking information (cf. Figure 7.4), a BOT can be safely accepted. If at least one resource is marked as being locked by another client, the current client can definitively reject a BOT (or a lockSubTree request). Only if resources are requested in a BOT for which no information is locally available, does the server have to be consulted.

7.3.3 What Does Locking a Concept Mean?

Locking resources in relational databases means the database administrators or application developers must decide whether to lock an attribute, a tuple, or a complete table (i.e. relation). Since the basic data model for ontologies is much richer (especially due to hierarchical relationships between concepts, between relations, and between instances and concepts) the decision of what a lock entails is more complex. The most simple answer would be to lock the complete ontology with all its components. But this solution is ruled out since it would disallow any kind of concurrency and distributed collaboration. Another simple answer would be to lock the resources that are to be modified within a transaction; for example, the resource X in the transaction that states that concept X has a super-concept Y. Apparently, for this transaction, concept Y should also be locked since a new sub-concept for Y is defined. Thus, the second simple approach seems to lock too few resources.

Due to hierarchical relationships between concepts locking a concept X implies *read-locks* for all super-concepts of X and all their super-concepts, recursively. A read-lock marks a resource as being read-only, that is, modifications to it are currently disallowed. If a read-lock for at least one super-concept cannot be achieved X will not be locked and the BOT fails. Thus, no operations may modify X. Read-locks can be available to multiple clients at the same time without conflict. If a client is the only one that read-locked a resource, the client can achieve a stricter (write) lock. Other clients cannot.

The reason why a lock propagates from one resource to another in the ontology can be seen in the following example scenario. Assume, X is a sub-concept of Y and Y has a slot A with range Y. Assume, we want to restrict the value range of A for X from Y to X. Thus, in the BOT we just lock the concept X and call the appropriate operation on X. Before we send the commit another client (after locking Y) changes the name of A to B and commits. If we now commit our transaction, the semantics of the combined operations is not defined. Does X now have two independent attributes A and B? Or is attribute A totally lost as well as our newly defined range restriction? Both situations are unsatisfactory. Thus, to prevent them, super-concepts need to be read-locked.

The same holds for locking complete sub-trees of the concept hierarchy. Here all sub-concepts are locked in the same way as the root of the sub-tree and all super-concepts of the root. All super-concepts of the sub-concepts of the root must be read-locked. This is necessary only if multiple inheritance is allowed. Because the same rules for computing super- and sub-objects of concepts, etc. are available in the client, in the server some decisions whether a transaction is allowed or not may be made on the client side without connecting to the server. Thus, the amount of queries sent over the network is reduced and processing times are enhanced.

7.4 Evaluation Phase

Once we have produced a target ontology, we need to evaluate it. The goal of the evaluation phase is to check whether the ontology fulfils the requirements specified during the first stage of the methodology. OntoEdit tackles several aspects for evaluation, that is, (i) test sets of instances and axioms can be used for the analysis of typical queries, (ii) a graphical axiom editor in combination with an underlying inference engine[6] allows for error avoidance and location, (iii) competency questions might be formalized into queries and evaluated by using the facilities of (i) and (ii) and, last but not least, (iv) a namespace mechanism allows the facilities (i)–(iii) to be used collaboratively.

7.4.1 Analysis of Typical Queries

The ontology engineer may interactively construct and save instances and axioms. OntoEdit contains a simple instance editor (cf. Figure 7.6) and an axiom editor that the ontology engineer can use to create test sets. A test set can be automatically processed and checked for consistency. Once the ontology evolves and needs changes to remain up-to-date, a test set may be re-used for checking the validity of the ontology. This basic functionality is needed, for example, during the use of competency questions. The ontology as well as related instances can be exported through export plug-ins to RDF(S), DAML+OIL in a file or be transferred directly into Sesame (the same holds for imports).

---

[6] The underlying inference engine used for processing of axioms is Ontobroker.

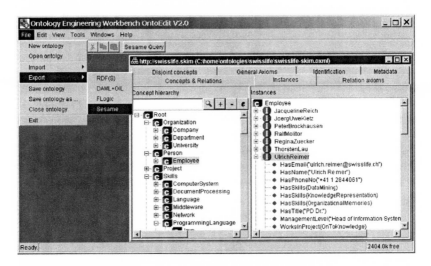

Figure 7.6 Instance editor/export plug-ins

7.4.2 Error Avoidance and Location

While the generation and validation of test cases allows for detection of errors, it does not really support the localization of errors. The set of all axioms, class and instance definitions sometimes express complex relationships and axioms often interact with other axioms when processed. Thus, it is frequently very difficult to overview the correctness of a set of axioms and detect the faulty ones. In order to avoid problems, OntoEdit allows for defining several standardized properties of relationships by clicking on the GUI (viz. symmetry, transitivity and inverseness of relations) and a graphical rule editor for other types of axioms. In order to locate problems, OntoEdit takes advantage of the underlying inference engine, which allows for introspection and also comes with a debugger. A very simple but effective method to test axioms with test cases, for example, is to switch off and switch on axioms. A more sophisticated approach uses visualizations of proof trees by tracking back the drawn inferences to the test instances. Therefore semantic errors in rules may be discovered. A more detailed description of OntoEdit's facilities for the 'analysis of typical queries' and 'error avoidance and location' can be found in Sure et al. (2002).

7.4.3 Usage of Competency Questions

Competency questions may help evaluation in two ways. First, they might provide prototypical instances for a test set (see above). Second, CQs define requirements for the ontology (ontology-based application), therefore they provide a checklist of questions the ontology should be able to answer. For example, from the CQ 'Who is head of a research group?' the concept group and the property headOfGroup (with domain Researcher and range group) are identified as relevant elements and therefore modelled in the ontology. A prototypical instance, for example, is an instance of Researcher, viz. 'Rudi Studer', who is headOfGroup of an instance of group, viz. the 'Knowledge Management Group'. Each CQ may now be formalized with the facilities described above into a query which is executed by the inference engine. The query result may be used to check whether the requirements expressed by the CQs are fulfilled by the current ontology.

7.4.4 Collaborative Evaluation

Taken together, the three facilities above can be used collaboratively through support from the backbone inference engine for the handling of multiple test sets. A namespace mechanism allows for syntactically splitting up ontologies or ontology parts (i.e. concepts, relations, instances and axioms) into modules that can be processed by a single instance or separate instances of the inference engine. Members of the engineering team usually have different requirements and use case scenarios, for example, expressed by their different CQs, therefore they typically need separate test sets for evaluation. In a two step approach, we (i) evaluate each test set locally, that is, each member (or each pair of ontology engineer and domain expert) evaluates his CQs, and (ii) evaluate globally the union of test sets.

7.5 Related Work

A good overview, viz. a comparative study of existing tools up to 1999, is given in Duineveld et al. (1999). Typically the internal knowledge model of ontology engineering environments is capable of deriving is-a hierarchies of concepts and attached relations. On top of that, our system provides facilities for axiom modelling and debugging. Naturally, the study does not fully consider the more recent developments, for example, Protégé (Fridman et al., 2000) and WebODE (Arprez et al., 2001).

    WebODE has a well-known methodological backbone, viz. METHON-TOLOGY, and is designed to integrate numerous aspects of an ontology life-cycle. Arprez et al. (2001) mentions that it offers inferencing services (developed in Prolog) and an axiom manager (providing functionalities such as an axiom library, axiom patterns and axiom parsing and verification), but the very brief mentioning of these functionalities is too short to assess precisely. About collaboration, it is said that this is supported at the knowledge level, but how this is achieved remains open.

Environments like *Protégé* (Fridman et al., 2000) or *Chimaera* (McGuinness et al., 2000) offer sophisticated support for ontology engineering and merging of ontologies. Protégé also has a modular plug-in design rational like OntoEdit, but lacks sophisticated support for collaborative engineering. They provide limited methodological and collaborative support for ontology engineering.

A system well-known for its reasoning support is *OilEd* (Bechhofer et al., 2001b) in combination with the description logics (DL) reasoner *FaCT* (Bechhofer et al., 2001a). Their collaborative and methodological support is rather weak.

Some tools explicitly support collaboration during ontology engineering. *APECKS* (Tennison and Shadbolt, 1998) is targeted mainly for use by domain experts, possibly in the absence of a knowledge engineer, and its aim is to foster and support the debate about domain ontologies. It does not enforce consistency or correctness, and instead allows different conceptualisations of a domain to coexist. *Tadzebao* (Domingue, 1998) supports argument between users on the ontology design, using text, GIF images and even hand-drawn sketches. The strength of these approaches lies in the advanced support for communication. In contrast, we provide a more sophisticated support for fine-granular locking of the ontology.

The web-based *Ontosaurus* (Swartout et al., 1996) combines support for collaboration with reasoning and allows individuals to expand an ontology only when consistency is retained within the ontology as a whole. This approach takes advantage of the reasoning and consistency checking abilities of the underlying representation language LOOM. Ontosaurus was inspired by the *Ontolingua* system (Farquhar et al., 1996), which does not have an inferencing support as an integral part of the ontology development environment. Due to the simple 'state-less' HL interaction, both systems have several limitations. For example, a server does not maintain any state information about users, that is, clients. Nor is it possible for a server to initiate an interaction on its own, for example, alerting users to simultaneous changes by others. In general, no other approach is known to us that implements fine-granular locking of ontologies like we do.

7.6 Conclusion

In this chapter we have presented the advanced collaborative tool support of OntoEdit for three major steps of the methodology for ontology development.

We have illustrated the features by examples from the Swiss Life case study on skills management. OntoEdit also has some features that could not be presented here, for example, an extremely capable plug-in structure, a lexicon component, and an ontology mapping plug-in. We have already applied OntoEdit and the methodology in additional scenarios (e.g. Mädche et al., 2002). OntoEdit has proven to be a flexible and powerful tool for developing ontologies.

For the future, it is planned to develop OntoEdit in several directions: (i) the collaborative facilities will be further expanded, for example, by adding a rights- and user-management layer on top of the locking mechanism, integrating communication and workgroup facilities and providing connectors to additional ontology servers and repositories; especially we want to implement a connection to the Ontology Middleware and Reasoning component (cf. Chapter 11) and Sesame (cf. Chapter 5); (ii) new imports and exports will be developed; (iii) the integration of ontology construction with requirement specification documents will be generalized by means of semantic document annotation; and (iv) the mind map tool will be more tightly integrated into the ontology engineering process, for example, through enabling direct communication between the tool and an ontology server, to name but a few.

8

QuizRDF: Search Technology for the Semantic Web

John Davies, Richard Weeks and Uwe Krohn

8.1 Introduction

Important information is often scattered across Web and/or intranet resources. Traditional search engines return ranked retrieval lists that offer little or no information on the semantic relationships among documents. Knowledge workers spend a substantial amount of their time browsing and reading to find out how documents are related to one another and where each falls into the overall structure of the problem domain. Yet only when knowledge workers begin to locate the similarities and differences among pieces of information do they move into an essential part of their work: building relationships to create new knowledge.

Information retrieval traditionally focuses on the relationship between a given query (or user profile) and the information store. On the other hand, exploitation of interrelationships between selected pieces of information (which can be facilitated by the use of ontologies) can put otherwise isolated information into a meaningful context. The implicit structures so revealed help users use and manage information more efficiently (Shipman et al., 1995).

Knowledge management tools are needed that integrate the resources dispersed across Web resources into a coherent corpus of interrelated information. Previous research in information integration (see, e.g. Hearst, 1998) has largely focused on integrating heterogeneous databases and knowledge bases, which represent information in a highly structured way, often by means of formal languages. In contrast, the Web consists to a large extent of unstructured or semi-structured natural language texts.

As we have seen, ontologies offer an alternative way to cope with heterogeneous representations of Web resources. The domain model implicit in an ontology can be taken as a unifying structure for giving information a common representation and semantics. Once such a unifying structure exists, it can be exploited to improve browsing and retrieval performance in information access tools. QuizRDF is an example of such a tool.

We now motivate the design of QuizRDF, a search engine that uniquely combines free-text search with a capability to exploit RDF meta-data in searching and browsing. There are three primary reasons for this approach, two of which are based on theoretical observations, and one of which is more pragmatic.

Pragmatically speaking, it is the case at the time of writing that only a very small proportion of WWW- and intranet-based information resources are annotated with RDF (meta)data. It is therefore preferable to provide a combined search facility that can exploit meta-data annotations where they exist but which will degrade gracefully to a 'traditional' free text search engine where information is not annotated.

Turning to more principled reasons for our approach, our own early user trials and, more importantly, a wealth of data from the information retrieval literature indicate that information seeking activity is often comprised of a mixture of searching and browsing behaviours. Research on user behaviour has typically characterized a variety of paradigms for information seeking and Bates (1986), for example, provides a discussion of these paradigms and their interactions. Similarly, Manber et al. (1997) describe a system for combining browsing and searching of WWW resources and argue that this combination delivers a more powerful tool for information seeking than search or browse facilities alone.

More specifically to the discussion at hand, browsing a graphical display of a comprehensive ontology can result in a high cognitive overload for the user. Instead of navigating the entire graph of a complex ontology, users may benefit more by starting from a particular node of the ontology and exploring its immediate surroundings in order to create appropriate queries. This raises the question of how to enable the user to find an interesting node in the ontology from which to start his exploration. In QuizRDF, as we will see, this is achieved by user entry of a 'standard' keyword search query which is used to locate them at an appropriate point in the information space represented by the ontology.

Another important observation is that it is in the general case impossible and impractical to cover the content of a document exhaustively by an RDF description. In practice, RDF descriptions can never replace the original document's content: any given RDF description of a set of resources will inevitably give one particular perspective on the information described. Essentially, a meta-data description can never be complete since all possible uses for or perspectives on data can never be enumerated in advance. Searches restricted to RDF descriptions will tend to produce a lower recall, while it is especially important at the beginning of a retrieval session to achieve a high recall. Most users are not able to initiate their search by formulating a complex and precise query. They prefer to start with a very simple query consisting of only one or two search terms in order to get a first idea of what information is available. Users may then continue their search by refining the queries to narrow the search results down to relevant documents.

Searching the full text of documents along with any associated RDF annotations can ensure the high recall desirable in the early stages of the information seeking process. In later stages of the search, when the user may typically be more interested in the precision of the retrieval results, it can be advantageous to put more emphasis on searching the RDF annotations.

QuizRDF can be used like a conventional Internet search engine by entering a set of search terms or a natural language query and produces a ranked list of links to relevant Web pages based on statistical algorithms (Salton, 1989) in the usual way. However, QuizRDF's indexing and retrieval technique is also designed to use domain knowledge that is made available in the form of ontologies specified as RDF Schemas. In our data model, RDF(S) is used to specify the classes in the ontology and their properties. The information items processed by QuizRDF are then RDF resources, which may be Web pages or parts thereof. Ontologically speaking, these RDF resources (WWW pages or parts thereof) are thus instances of the classes defined in RDF(S).

In the next section, we describe in detail the indexing process used by QuizRDF and then proceed to discuss in Section 8.3 how this index is exploited in the retrieval process. Finally, we briefly describe our experience to date in using QuizRDF and some avenues for further research.

8.2 Ontological Indexing

QuizRDF's indexing and retrieval technique is designed to use domain knowledge that is made available in the form of ontologies specified as RDF Schemas. The information items processed by QuizRDF are generally RDF resources, which may be whole Web pages or parts of Web pages. QuizRDF uses a given RDF Schema to create a structured index of RDF resources.

The core of the indexing process in QuizRDF is the assignation of content descriptors to RDF resources (Web pages or parts thereof). Content descriptors of a resource are terms (words and phrases) that QuizRDF obtains from both a full text analysis of the resource content *and* from processing all literal values that are directly related to the resource by a property (recall that an RDF resource is an instance of a class in the ontology). The QuizRDF index also retains structural information about the ontology from the corresponding RDF(S) description (e.g. classes, their properties and the sub/super-class relations holding between them).

In practice, the ontological index created by QuizRDF is a set of triples that refer to a set of RDF resources in a manner analogous to the way in which constituent terms (words and phrases) refer to documents in 'traditional' information retrieval systems (Salton, 1989).

In the RDF meta-data, URLs are instances of classes, as defined by the `typeOf` property and we can write this in our scheme as:

$$<URL_n, \text{typeOf}, \text{Employee}>$$

Furthermore, values of properties can be written down as follows:

$$<URL_n, \text{last\_name}, \text{"Miller"}>$$

QuizRDF creates a multidimensional index by combining such triples as follows:

$$<\text{"Miller"}, \text{Employee}, \text{last\_name}> \rightarrow URL_n$$

which represents the fact that the resource at URL_n is an instance of class *employee* and that this instance's *lastname* property has value "Miller". More generally, a set of triples of the following type are produced:

$$<\text{literal}, \text{class}, \text{property}> \rightarrow URL$$

At the same time the full text of the annotated URLs is indexed in the way familiar from the information retrieval literature (conceptually at least) creating further triples:

$$<\text{"George Miller"}, \text{Employee}, \phi> \rightarrow URL_n$$

So the triple above represents the fact that the phrase "George Miller" occurs in the body of the document at URL_n and that this document is of type *Employee*.

Figure 8.1 shows a simple example to illustrate ontology-based indexing. When indexing the example Web page at malta.bt.com/gm/cv, QuizRDF not only analyses the full text of the resource content but also the relevant parts of the RDF graph that describe this resource.

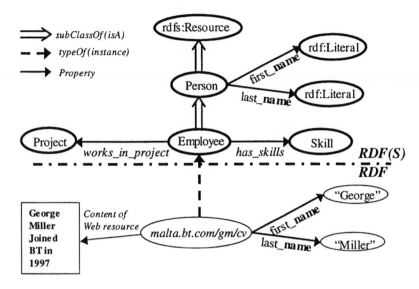

Figure 8.1   Ontology-based indexing

Since our example Web page is annotated as being of type *Employee*, QuizRDF processes all literal values that are directly related to the class *Employee*. The properties connected to *Employee* in this example are *last_-name*, *first_name*, *has_skills* and *works_in_project* (*last_name* and *first_name* being inherited from the class *Person*). QuizRDF traverses the RDF graph along these properties and performs a full text analysis on those properties having literal values. (Note that QuizRDF will also convert numeric literal values to strings for similar indexing). The resulting index contains descriptors extracted from both the full text content of the Web page and the RDF graph.

The content descriptors obtained are stored in QuizRDF's index along with references to their structural origin. Figure 8.2 illustrates schemati-

Descriptor	Class	Property	Resource
Miller	Employee	Ø	malta.bt.com/gm/cv
joined	Employee	Ø	malta.bt.com/gm/cv
BT	Employee	Ø	malta.bt.com/gm/cv
1990	Employee	Ø	malta.bt.com/gm/cv
George	Employee	first_name	malta.bt.com/gm/cv
Miller	Employee	last_name	malta.bt.com/gm/cv

Figure 8.2   Ontology-based index

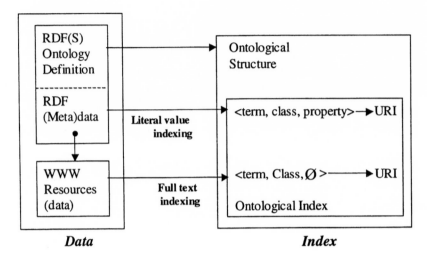

Figure 8.3   Ontological index structure

cally the data structure used to store the index for the given example. For each extracted keyword the index maintains a reference to the resource type *Employee*. For descriptors obtained from literal values, the database additionally stores the name of property that relates the literal to the given instance of *Employee*.

Figure 8.3 gives a more general overview of the full QuizRDF index structure.

8.3 Ontological Searching

In this section, we describe how the ontological index above is used to provide a natural and intuitive browsing and searching interface onto a set of RDF-annotated WWW information resources.

On start-up, QuizRDF presents the user with a text entry box and a drop-down menu. The drop-down menu contains a list of all the resource types stored in the QuizRDF index. The user can enter any natural language text into the text entry box. QuizRDF responds by returning a list of RDF resources ranked according to a resource's relevance to the user query (the ranking is currently based on a variation of the well-known *tf.idf* vector product scheme (Harman et al., 1992). Simultaneously, the classes of which the URLs in the results list are instances are computed and included in the drop-down list above the results list. Selecting a class then (i) filters the retrieval list to include only those URLs which are instances of the selected class and (ii) displays the properties and related classes to the selected class, each of which has a hyperlink associated with it allowing the user to browse the ontology.

For each attribute the user can input a search criterion. QuizRDF combines the search criteria entered (which can be both free text search terms and attribute values) and matches the resulting query against its ontology-based index. In addition, resource types (classes) related by some property to the currently selected type are displayed as hyperlinks. Clicking on such a type then selects that type and in turn displays those types which are related to it. Thus, the user can browse the ontology in a natural and intuitive way.

To exemplify, Figure 8.4 shows a typical initial query by a user using the skills ontology described above. The user has entered a free text query for information about an employee called George Miller. The search engine has

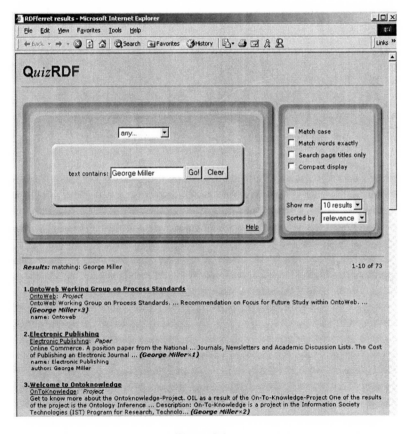

Figure 8.4

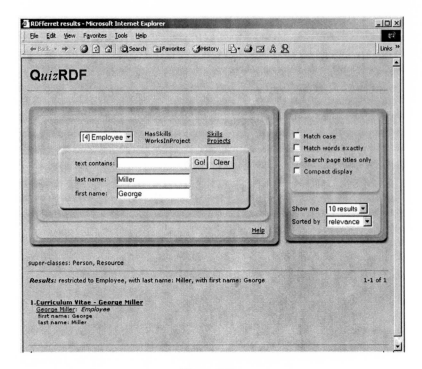

Figure 8.5

returned a ranked list of 73 documents mentioning the terms 'George' and/or 'Miller'. At the top of the screenshot can be seen a drop-down list containing the selection 'any…'. When returning the 73 results documents, QuizRDF has also compiled a list of the classes to which each document belongs. This class list is then made available to the user via the drop-down list referred to.

Figure 8.5 shows the result after the user has selected the *Employee* class from the drop-down list. The screen now shows the properties of the *Employee* class, differentiating between those which relate (instances of) this class to (instances of) another class (*HasSkills* and *WorksInProject*) and those which expect a literal value (*last name* and *first name*). The user has then specified values (Miller and George, respectively) for these properties, while now leaving the free text search box empty. Based on these selections, QuizRDF has identified in its results list the single document (instance) of class *Employee* fulfilling the criteria specified. The super-classes of Employee (*Person* and *Resource*) are also identified. These super-classes, as well as the classes linked to *Employee* by properties (*Skills* and *Projects*), are clickable, allowing the user to continue to browse the ontology, initiating new searches at any point.

8.4 Alternative data models

The data model used above to exemplify the use of QuizRDF sometimes turns out in practice to be a little simplistic. It will be recalled that a Web resource in the model was an instance of an ontological class (*Employee*, say). Intuitively, one interpretation of this is that the Web resource is 'about' in some sense a particular employee (George Miller in our examples above). For some information sets, however, the Web resources may typically be 'about' more than one ontological class. Consider, for example, a Web page which has a section describing the life of Van Gogh and a further section with reproductions and descriptions of some of his most famous paintings. Suppose further that the ontology at hand has classes 'painter' and 'painting': of which class are we to say the given page is an instance? To choose only one of the two valid classes seems to throw away some valid domain knowledge that could be captured were we to allow the page somehow to be 'about' both relevant classes.

An alternative data model has thus been developed to deal with this problem. The new model effectively introduces a 'level of indirection'. Specifically, each instance of a class is now a URI which does not refer to a WWW resource. Instead, these new URIs are related to URLs via the is-about property. Crucially, this property is n:1 (i.e. many URIs can point to 1 URL, allowing any given Web resource to be 'about' many of the new URIs).

8.4.1 Indexing in the New Model

URIs are now instances of classes, as defined by the type-of property:

    <URI_n, typeOf, Employee>

Values of properties are defined as follows:

    <URI_n, last_name, "Miller">

There is now a new type of triple in the index:

    <URL_n, is-about, URI_n>

QuizRDF now creates a 'multidimensional index' from such triples as follows:

    <"Miller", Employee, last_name> → URI
    <literal, class, property> → URI

At the same time the URLs are full text indexed as before, creating further triples:

    <"George", Employee, ϕ> → URI

8.4.2 Searching in the New Model

Now when a user enters a query, a ranked list of URLs is returned in the usual way, based on the terms (literal values and words) referencing the URIs in the index. However, the referenced URIs may of course point to the same URL. In this instance, two or more URIs are combined in the results list and link to the same URL. Where labels for the URIs are provided in the RDF descriptions, these are used in the results list as the name(s) of a given URL; where they do not exist, the URL string itself is listed.

8.5 Further Work

8.5.1 Technical Enhancements

A number of possible enhancements to QuizRDF are in progress or under consideration and we briefly mention three of them here.

As we have seen, the resources returned from QuizRDF can be clustered based on their types (i.e. classes of which they are an instance). The proposal is that each resulting cluster can then be 'scored' by combining some score of the individual WWW resources contained therein. Currently, as mentioned above, QuizRDF uses a variation of the well-understood *tf.idf* scoring scheme to rank WWW resources against a user query (Harman et al., 1992). In this way, each resource type (ontological class) can be ranked by relevance to the user query and results can be presented aggregated around resource types. Currently, we are developing a text-based way of presenting these results but there are clearly opportunities to exploit a graphical interface here also.

A second area of ongoing work stems from an obvious limitation of QuizRDF as currently implemented: namely, queries can only be made around one class. To exemplify, we can ask the queries

'Find me all employees with last name Miller' and
'Show me all instances of the class painter'

but we cannot (at least not in a single step) ask queries involving any 'chaining', for example:

'Find me all instances of class painting painted by an (instance of class) painter whose first name is Pablo'

Allowing this possibility without comprising the simple and intuitive interface of QuizRDF is the subject of ongoing research.

The third area of current work involves the implications of indexing large ontologies in QuizRDF. The issue is not one of scalability at the indexing level: the technology underlying QuizRDF has been used to comfortably index a collection of 3.5 million documents. Rather, the issue arises at the user interface level: in a heavily interconnected ontology, or one in which classes typically have a large number of properties, how can we display many classes and properties intuitively in the QuizRDF interface? Some initial work has been carried out involving use of drop-down lists rather than lists of hyperlinks but further effort is needed.

8.5.2 Evaluation

In addition to BT itself, several other organizations are interested in evaluating QuizRDF. We briefly mention here the use that EnerSearch intend to make of the system. EnerSearch is a virtual organization researching new IT-based business strategies and customer services in deregulated energy markets on behalf of its member companies. As such, EnerSearch is a knowledge creation company. In common with most WWW-based information, EnerSearch's WWW site for the use of its members holds weakly structured information in mixed media. EnerSearch's site is a key component for delivery of its primary role: the transfer of knowledge to its shareholders (member companies), employees and other interested parties. It is intended to carry out a study to compare the effectiveness of EnerSearch's current 'text-only' search engine with the combination of free text search and structured ontological browsing which is embodied in QuizRDF.

As described in Chapter 13, EnerSearch have designed a detailed experiment to evaluate the advantages of QuizRDF, using both qualitative and quantitative evaluation techniques.

8.6 Concluding Remarks

Discussion with potential users as well as evidence from the information retrieval literature indicated clearly the desirability of combining RDF browsing and querying with full text search. A full text search capability means a user can enter a relatively simple initial query that essentially quickly locates them in the information space, from where further browsing and searching can proceed. Additionally, supporting full text search means that the user can access the information even at an early stage when annotations are still sparse, while the support for RDF in QuizRDF allows structured browsing of an ontology. As RDF annotations are added to the system the user will benefit from the high precision and semantic expressiveness of RDF querying. This can be seen as a *low threshold, high ceiling* approach: the user can start using QuizRDF without necessarily having to invest a lot of time in creating a rich set of annotations; while on the other hand, every newly added annotation will have an immediate effect on the system's performance and usability.

We have argued that QuizRDF's combination of ontological browsing and free text search supports a more natural and intuitive information seeking process than is available in either a search engine or a browsing tool alone.

We have described our initial implementation of QuizRDF and indicated some further directions of research and briefly discussed an ongoing evaluation of the system. QuizRDF is an early example of the much-improved information access tools that the advent of the Semantic Web makes possible.

9

Spectacle

Christiaan Fluit, Herko ter Horst, Jos van der Meer,
Marta Sabou and Peter Mika

9.1 Introduction

Many Semantic Web initiatives improve the capabilities of machines to
exchange the meaning of information with other *machines*. These efforts
lead to an increased quality of the application's results, but their user interfaces take little or no advantage of the semantic richness. For example, an
ontology-based search engine will use its ontology when evaluating the user's
query (e.g. for query formulation, disambiguation or evaluation), but fails to
use it to significantly enrich the presentation of the results to a human user. For
example, one could imagine replacing the endless list of hits with a structured
presentation based on the semantic properties of the hits.

Another problem is that the modelling of a domain is done from a single
perspective (most often that of the information provider). Therefore, presentation based on the resulting ontology is unlikely to satisfy the needs of all the
different types of users of the information. So even assuming an ontology for
the domain is in place, mapping that ontology to the needs of individual users
– based on their tasks, expertise and personal preferences – is not trivial.

9.2 Spectacle Content Presentation Platform

The Spectacle Semantic Web initiative improves the capabilities of machines
to present the meaning of information to *humans*. Spectacle is a content
presentation platform featuring custom-made information presentations,
aimed at supporting the information needs of its users. This means not only

that the right information should be delivered to the user, but also that it needs to be presented (structured, formatted, rendered) in a manner appropriate for that specific user.

Spectacle is used to disclose both the content of databases, document repositories and other enterprise information sources, as well as the semantics of that information from Semantic Web resources.

For the end user, Spectacle transforms the task of gathering information from a search task (formulating explicit queries) to a browsing task (using navigation heuristics) by presenting each user with the navigational means appropriate for his or her task. This results in more efficiency in retrieving the right information, both in terms of retrieval accuracy as well as time spent on the task.

Spectacle can present information in two different ways: firstly, it can create hypertext interfaces, containing selected content, design and an appropriate navigation structure, based on the semantics of the information. Additionally, it can present the information by graphical visualization.

A key benefit of the first approach is that it allows for an easy and flexible presentation of the same information in different ways, for each of the envisioned tasks or user groups. Furthermore, it has all the usual benefits of a generated web site (like having a consistent design, being up-to-date) and it also takes advantage of the expressivity and flexibility provided by Semantic Web standards such as RDF, RDF Schema and DAML+OIL.

A benefit of the second approach is that it can offer insights and kinds of information access that are not possible with conventional publishing methods such as web sites. For example, overview and analysis of large sets of objects requires an effective and compact graphical presentation. Similarly, presentation of the relations between these objects is virtually impossible without the support of a graphical visualization.

9.2.1 Ontologies in Spectacle

What is required is a powerful information organization metaphor that can still be understood by *all* users. The first candidate that comes to mind is the list, which fails to comply with the 'powerful' requirement. The next step up is a tree-like structure, which is significantly more powerful than a list, yet is still easily understood and widely used, for example in menus, directory trees, bookmarks, etc. Anything potentially more powerful, for example, unrestricted graphs, is too complex for the average user to handle.

For this reason, Spectacle maps the existing complex domain ontologies to lightweight ontologies that describe the domain through a set of classes (concepts) and their hierarchical relationships. Also known as taxonomies, this type of ontology is already being used in several domains (biology, chemistry, libraries) as classification systems. Information architects consider taxonomies to be basic building blocks, representing the backbone of most web sites. Non-formal taxonomies are already widely used in web applications for product classification (e.g. Amazon[1]) or web directories (e.g. Yahoo,[2] Open Directory Project[3]). Taxonomies are also part of Semantic Web standards such as RDF and Topic Maps.

Due to the specialization relationship that is encoded in the hierarchy, the set of objects in a subclass is a subset of the objects of its super-class. The set of subclasses of a class is called *complete* when their union contains all the objects of the super-class. Classes that share instances are called *overlapping* if no specialization relationship holds between them.

These characteristics are very common for taxonomies. Especially the 'overlapping' characteristic of a taxonomy proves very valuable, because it provides multiple routes to the same information (when used in navigation) and insight into correlations between concepts (when used for graphical analysis).

9.3 Spectacle Architecture

For the transformation of information from existing sources to an ontology-based presentation, Spectacle offers an application development architecture consisting of the following components (Figure 9.1):

- *Spectacle Onto API.* Supports the ontology mapping introduced in Section 9.4.2. A detailed description of the mapping process is provided in Section 9.4.
- *Spectacle Core API.* Provides an interface to the Spectacle server. This API is used internally by the Onto API and is not discussed here.
- *Spectacle Visualization API.* Provides the means to generate and interact with the visualizations described in Section 9.5.

9.4 Ontology-based Mapping Methodology

Applications based on the Spectacle Onto API create a transient ontology-based mapping of data sources to a Spectacle presentation. The Onto API provides a framework for application development by offering a controlled procedure for the transformation process. It supplies the application developer

[1] http://www.amazon.com/
[2] http://www.yahoo.com/
[3] http://www.dmoz.org/

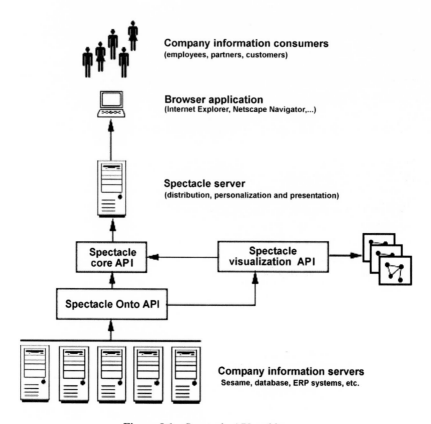

Figure 9.1   Spectacle API architecture

with domain-independent methods and components that facilitate the development.

The procedure consists of the following steps:

1. Identify relevant information entity types in the data sources and the data to display for each entity.
   For each distinct entity type:
2. Create a mapping of terms, relations, properties and property values from the domain ontology to taxonomical terms.
3. Decide on how to render the individual entities in the presentation.
4. Decide on the navigation paths through the ontology that are relevant to the entire entity set, based on the properties identified in step 2.
5. Decide on which information to render in each step of the navigation sequence.

6. Decide on filters that limit the entire set on certain criteria, yet use one of the navigation sequences identified in step 4.
7. Determine typical user profiles, based on the information views from step 4 and 6.

Example. *Throughout this chapter, we will use examples from the Ener-Search case study. EnerSearch is a virtual research organization that uses Spectacle in order to provide access to its extensive collection of publications through ontology based navigation. The ontology of EnerSearch is a combination of a lightweight ontology that describes the domain of EnerSearch and a rich ontology that contains meta-data such as the author and title of publications. For further information, please consult Chapter 13.*

9.4.1 Information Entities

The first step in the Spectacle Onto scenario is to identify relevant information entity types in the data sources. Entities are real-world concepts such as 'products' or 'documents', information about which is stored in – or can be derived from – the data sources.

Example. *EnerSearch identified the need to offer the users navigation on the level of pages, as opposed to entire publications, for example. This level of granularity is made possible by named anchors in the HTML pages that can be used to point to specific parts within publications.*

9.4.2 Ontology Mapping

In this step, the taxonomical terms are determined by which the entities are classified, based on the characteristics of the available content. The classification terms may correspond to types from the domain ontology, property values, ranges of property values, combinations of property values, or relations between instances. These terms will be used later for navigation and selection purposes. The key benefit of this step is that an abstraction is made (for the purpose of presentation and navigation) from the arbitrary modelling decisions made in the domain ontology.

Example. *The EnerSearch presentation employs several classifiers. Entities (i.e. pages) are classified by individual concepts, authors, projects and the year of publication. Concepts are classes in the ontology, while authors and*

http://143.217.135.12/evaluation/knowledgebase/publications/conference-journals/ismick96/IBMICK96-3.html#page_11

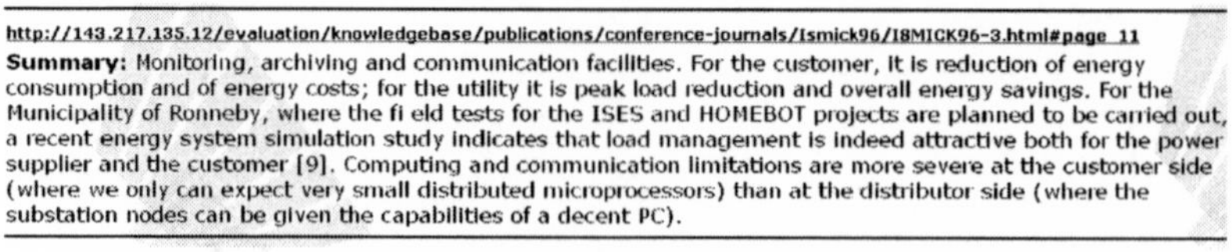

Summary: Monitoring, archiving and communication facilities. For the customer, it is reduction of energy consumption and of energy costs; for the utility it is peak load reduction and overall energy savings. For the Municipality of Ronneby, where the fi eld tests for the ISES and HOMEBOT projects are planned to be carried out, a recent energy system simulation study indicates that load management is indeed attractive both for the power supplier and the customer [9]. Computing and communication limitations are more severe at the customer side (where we only can expect very small distributed microprocessors) than at the distributor side (where the substation nodes can be given the capabilities of a decent PC).

Figure 9.2   Example entity rendering

projects are property values. Classification by year is an example of classifi-cation by a range of property values, since a year is defined as a range of date values.

9.4.3 Entity Rendering

In this step, the default rendering of each entity type is determined. Typically, some form of HTML or XML template is chosen or developed in which the relevant information for each entity can be inserted. Note that this default rendering can later be overridden for specific user profiles (Figure 9.2).

Example.   *Entities are shown as a link along with a summary of the content, based on an HTML template. By clicking on the URL the content opens in a separate window for viewing. The source of the information that is used for rendering the entities is the ontological meta-data.*

9.4.4 Navigation Specification

In this step, the sequence of taxonomical terms is determined by defining generally applicable paths by which users can navigate through the taxonomy. Each level in the navigation is expressed as the set of values for a term obtained in Section 9.4.2. Each sequence of levels leads to a simple navigation specification. Multiple navigation specifications are possible for each set of entities.

In order to allow the end user to select the next most relevant taxonomical term (instead of just a value of that term), the concept of a meta navigation step was introduced. A meta navigation step differs from the regular (content) navigation step, in that it allows the user to select a taxonomical *term*, instead of one of the values of that term. Note that a meta navigation step does not actually narrow down the selection for the user. For that reason, a meta navigation step is normally combined with the corresponding content naviga-tion steps in the user interface.

Example.   *In the EnerSearch case, there is only one navigation specification* *that is defined in terms of concepts. However, only key concepts are included on the top level of the navigation, as it is expected that users would naturally start navigating by the most important concepts.*

9.4.5 Navigation Rendering

In this step, a decision is made as to what to render in each step of the navigation obtained in the previous section (and how to render it). For example, it might not be deemed appropriate to show the default entity rendering at each level in the navigation, because this would lead to information overload for the user. So higher levels in the navigation hierarchy commonly use overviews (showing only a summarized rendering of a limited set of matching entities, and only when the user has drilled down to a sufficiently small set of information are the full entities shown.

An important aspect of the navigation rendering is that navigation paths are only offered when they lead to a non-empty set of entities. The user will never be confronted with a navigation step that results in an empty page.

Example.   *The topmost level is shown as a menu in a separate frame in the EnerSearch presentation while all other navigation steps are rendered using*

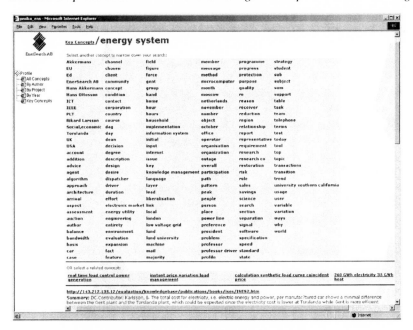

Figure 9.3   Example navigation rendering

in-page navigation. In other words, the links that the user can use to navigate further are embedded within the page. See Figure 9.3 for an example.

On the top of each section that is rendered from a navigation step, there is a table that contains all concepts that occur in the current selection of pages. Below, in a separate table, the concepts related to the previously selected concept(s) are shown. These provide an alternative way to continue the navigation. At the bottom the entities in the current selection are rendered as discussed above.

The presentation also makes use of configurable heuristics built into Spectacle that help to reduce the number of sections that needs to be rendered. Permutations of navigation steps are recognized, for example the step 'energy/saving' is equivalent to 'saving/energy' and therefore needs to be rendered only once. Furthermore, in each section the steps that do not cut back the selection significantly are not shown, as these choices lengthen the navigation path without adding much to the definition of the user's query. Also, nodes that contain less than a minimum number of entities are not extended any further, as it is expected that the user can handle a smaller selection without further help from the system.

9.4.6 Views

In this step, useful subsets of the entire set of entities that nonetheless use the same navigation are determined. Views are used instead of just introducing extra navigation steps in order to avoid an internal combinatorial explosion of the number of navigation paths. Views are integrated into the navigational structure by mounting them at a particular location in the navigation tree. Conceptually and visually, there is no difference between a view and the regular navigation tree; views are a tool to be able to control the combinatorial explosion that occurs when introducing navigation steps.

Example. *Navigations by the author, project or date of publication are implemented as views. For example, the view that is mounted at 'By Author/Akkermans, Hans' selects the subset of pages that belong to publications written by Hans Akkermans.*

9.4.7 User Profiles

In the final step it is determined which of the default navigation(s) and views are appropriate for the identified user groups. For example, end users may never get to see the full set of entities and have only one navigation path, while administrators may get access to all information along multiple navigation paths. User profiles may also control the rendering of the information, so that internal properties can be hidden from regular users. The user profiles decided upon here are just defaults, and can be extended and adapted by individual users if required.

Example. *Although the EnerSearch case study does not employ user profiles at the moment, these could be used to provide privileged access to internal publications for selected users, such as the shareholders of the company.*

9.5 Ontology-based Information Visualization

In the previous section, a textual, web-based rendering was used to present the information to the end user. This presentation is appropriate for the average information consumer due to its low entrance threshold: the user can find information by simply browsing through it. However, a textual presentation often hides certain characteristics of the information that are necessary for other tasks that require more insight into the structural properties of the information.

The Spectacle Cluster Map is an information visualization component that visualizes the objects of a number of selected terms from a taxonomy, using these terms as the primary organization principle. Cluster maps support at least three general information-seeking tasks: *analysis*, *querying* and *navigation*.

9.5.1 Analysis

A user may want to analyse a data set in order to get a better insight in its characteristics, to get a global understanding of the collection and to discover patterns and exceptional cases.

The cluster map in Figure 9.4 shows a collection of pages from the publications of EnerSearch, organized according to a select number of key concepts from the ontology. Each small yellow sphere represents a page. The larger, green spheres represent taxonomical terms. Directed edges indicate the hierarchical relationships between terms and point from specific to generic (e.g. *knowledge management* is a subclass of *management*). Balloon-shaped edges connect a set of objects to their most specific term(s). Objects that are classified by the same set of terms are grouped in *clusters*. Our example contains seven clusters; three of them represent overlaps between terms.

This visualization is immediately more expressive than its text-based counterpart with respect to the gained global insight into the taxonomy. The terms and their relationships are easy to detect. Also, it is immediately apparent which objects belong to one or multiple terms, which terms have shared

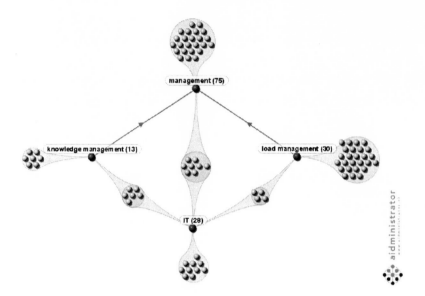

Figure 9.4   An example cluster map

instances (e.g. IT and knowledge management) and which do not (load management and knowledge management). Also it is clear that IT plays a part in both types of management. The relative sizes of the classes are clearly visible, e.g. there are a lot more pages written about load management than knowledge management.

Another interesting aspect of the visualization is that proximity in the map is related to semantic proximity. This is a consequence of the graph layout algorithm. *Classes* are semantically close if they share many instances. Indeed, the more instances two classes share, the closer they are in the visualization. *Instances* are semantically close if they belong to the same class(es). Indeed, instances that have the same class memberships are explicitly clustered.

Figure 9.5 shows a more extensive example of a cluster map, displaying the cluster model of four key terms within the EnerSearch domain. Here, the clusters are displayed as cylinders, abstracting from the individual instances they contain. This allows the visualization to scale to larger amounts of instances. Furthermore, the edges connecting the same class to various clusters all have the same colour, so that the distribution of its instances over the map is easier to spot.

The *static version* of the cluster map (i.e. the image itself) already contains a lot of information that can be useful for analysing the data set. The classes and their hierarchical relationships provide an understanding of the domain of the

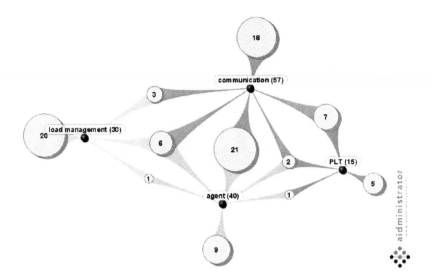

Figure 9.5   A cluster map showing countries vs. accommodation space

data set. The way instances are classified results in characteristics such as
incompleteness and overlaps, showing class relationships at the instance
level. The cardinality of classes and clusters supports a quantitative analysis.
By interpreting just this information, one can already come up with some
domain-specific observations.

Additionally, the interactive cluster map viewer, shown in Figure 9.6,
makes analysis an interactive and explorative process. The overview, zoom
and filter facilities offered by the user interface qualitatively enhance the
analysis process, by allowing the user to graphically browse through the
taxonomy.

The *strategy* used to obtain one or more cluster maps plays an important
role in its application. Different strategies can support different analysis
scenarios:

1. *Analysis within a single domain.* In this case a data set is visualized from
   one or more perspectives, giving insight into the collection. In one of our
   pilots we investigated the characteristics of a set of job offers by visualizing
   them according to the region or the relevant economic sector.
2. *Comparison of different data sets.* Data sets can be compared by visualizing
   them using the same ontology. For example, we have compared the offer-
   ings of two banks, by imposing the same taxonomy on their web sites and
   analysing the two visualizations.

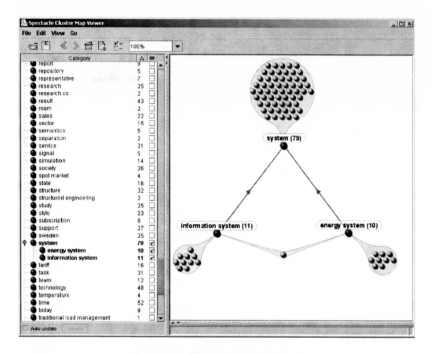

Figure 9.6   The cluster map viewer

3. *Monitoring.* Analysis of a data set at different points in time provides insight into the way it evolves. For example, one can monitor the web site of a company over time and see how its activities evolve.

9.5.2 Querying

The goal of a query task is to find a narrow set of items in a large collection that satisfy a well-understood information need (Marchionini, 1995). Query functionality for document collections is typically provided by a query form in which the user can enter query terms, optionally combined with Boolean operators. Several factors limit the effectiveness of this approach. Often, users find it difficult to come up with the right terms, because they only have a vague understanding of their information need and/or because they use a different vocabulary. Furthermore, they often have great difficulty formulating effective queries using Boolean expressions (Shneiderman, 1996).

The cluster map offers an alternative, graphical approach to querying. A user interface, such as shown in Figure 9.6, may allow the user to select a number of classes from the taxonomy, which become the query terms. Presenting the user with a collection of terms solves the vocabulary mismatch problem. Moreover, it will often lead to more precise queries since users may encounter additional applicable terms that they would not have thought of by themselves.

A cluster map displaying these classes and their instances then serves as a graphical representation of the hit list. For example, the left side of Figure 9.7 shows the result of the query '*load management OR research OR deregulation*'. The individual instances in the map represent the hits that match this query. A different colour scheme is used that shows the degree of relevance: the more relevant the hits, the darker the colour of the edges connecting them to their classes. Consequently, the best hits visually stand out from the map.

Clearly, there is no hit that satisfies all the requirements, since there is no overlap between *all* classes. This scenario is well known in electronic commerce: when searching for a product, a customer selects a set of values that are of interest. If the query does not fully succeed then either a 'no results' message or a long list of partial matches is shown. The customer gets neither a clear overview of the results nor suggestions for further exploration. As a result he often abandons the site. The ideal situation would be that the customer is guided to alternatives that would come close to his needs. With this visualization the customer has the opportunity to analyse alternative solutions in terms of his original query. For example, the map indicates that, although there is no instance satisfying all criteria, a hit is obtained when one of two requirements is dropped.

There is no need to formulate Boolean expressions, since they are an inherent part of the visualization. The following expressions are already expressed by the graph structure:

1. the union of the classes (disjunction of *all* query terms);
2. all intersections of the selected classes (conjunction of *some* query terms);
3. a special case of two: the intersection of all classes (conjunction of *all* query terms) – if any.

If the user wants a disjunction of the terms, he will analyse all the presented instances. As an added value he will see how the corresponding classes overlap, providing more insight into the resulting set. A more interesting (and probably more frequent) case is when a user wants the conjunction of the terms. In that scenario, two extreme situations can happen:

1. the result set is too large (under-specification);
2. the result set is empty (over-specification).

If the result set is empty, the user can still find instances that partially satisfy

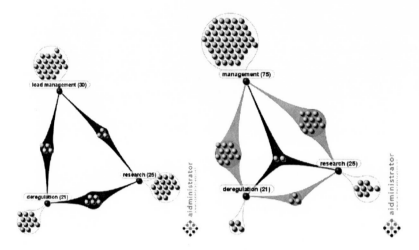

Figure 9.7   Two cluster maps showing the hit lists of two queries

the query, as shown above. The colour scheme assists in picking out the best hits. This is a form of *query relaxation*.

It is also possible to narrow or broaden the scope of the search by *refining* the query. If the result set is too large, the user can replace some classes with more specific subclasses. If the result set does not contain the item the user is looking for, he can try to broaden the search by replacing classes with their super-classes.

This principle is illustrated by the two cluster maps shown in Figure 9.7. While no single instance matches all terms of the query on the left, the right side shows that an optimal hit can be found by broadening the search to all forms of management. Note that both narrowing and broadening the scope of the query are possible due to the subclass relations in the taxonomy; therefore this capability is strongly dependent on a well-structured hierarchy.

Summarizing, we can say that this visualization has transformed the query process into a browsing process.

9.5.3 Navigation

Cluster maps can also be used for graphical navigation. We have employed them as image maps in Spectacle-generated web sites based on taxonomical data. Two navigation scenarios have been implemented, as described below.

In the *first scenario*, the cluster map is used in addition to the more traditional navigation facilities used in these sites: textual trees. It plays the role of a *site map* that can be invoked by the user when needed. It presents an overview of the whole data set: it shows the most important classes, their relationships and their instances. One can click on a class or a cluster of instances, which links to a corresponding location in the navigation structure. The role of the map is to facilitate a quick understanding of the available content and to provide quick access to individual items.

In the *second scenario,* the cluster map is always present as it has become the only navigation facility. Maps gradually present deeper levels of the ontology: the user starts with a visualization of the top of the taxonomy, showing only top-level classes, and can navigate towards more specific topics by clicking on a class of interest, which then becomes the *current class.* At any point, the map shows the current class and its super- and sub-classes (i.e. all classes that it has a direct hierarchical relationship with). For the current class, its instances are also presented in a textual list alongside the map. This hierarchical browsing facilitates a levelled understanding of the data.

9.6 Summary: Semantics-based Web Presentations

The technologies described in this chapter, supported by other Semantic Web technologies described in this book, enable organizations to exploit their semantically enhanced information resources to enrich the information presentation towards individual users. One presentation technique excels at low entrance threshold, easy to understand, browsable presentations, while the other offers highly expressive visualizations that provide the insight necessary for more complex tasks. Depending on specific needs, the right combination of these two techniques will provide optimal information access for knowledge workers.

10

OntoShare: Evolving Ontologies in a Knowledge Sharing System

John Davies, Alistair Duke and Audrius Stonkus

10.1 Introduction

We saw in the introduction how the Semantic Web makes possible a new generation of knowledge management tools. We now turn our attention more specifically to Semantic Web based support for virtual communities of practice. The notion of communities of practice (Seely-Brown and Duguid, 1991) has attracted much attention in the field of knowledge management. Communities of practice are groups within (or sometimes across) organizations who share a common set of information needs or problems. They are typically not a formal organizational unit but an informal network, each sharing in part a common agenda and shared interests or issues. In one example it was found that a lot of knowledge sharing among copier engineers took place through informal exchanges, often around a water cooler. As well as local, geographically based communities, trends towards flexible working and globalisation have led to interest in supporting dispersed communities using Internet technology (Davies, 2000a). The challenge for organizations is to support such communities and make them effective. Provided with an ontology meeting the needs of a particular community of practice, knowledge management tools can arrange knowledge assets into the predefined conceptual classes of the ontology, allowing more natural and intuitive access to knowledge.

Knowledge management tools must give users the ability to organize information into a controllable asset. Building an intranet-based store of information is not sufficient for knowledge management; the relationships within the stored information are vital. These relationships cover such diverse issues as relative importance, context, sequence, significance, causality and association. The potential for knowledge management tools is vast; not only can they make better use of the raw information already available, but they can sift, abstract and help to share new information, and present it to users in new and compelling ways

In this chapter, we describe the OntoShare system which facilitates and encourages the sharing of information between communities of practice within (or perhaps across) organizations and which encourages people – who may not previously have known of each other's existence in a large organization – to make contact where there are mutual concerns or interests. As users contribute information to the community, a knowledge resource annotated with meta-data is created. Ontologies defined using the resource description framework (RDF) and RDF Schema (RDFS) are used in this process. RDF (Lassila and Swick, 1999) is a W3C recommendation for the formulation of meta-data for WWW resources. RDF(S) (Brickley and Guha, 2000) extends this standard with the means to specify domain vocabulary and object structures – that is, concepts and the relationships that hold between them.

In the next section, we describe in detail the way in which OntoShare can be used to share and retrieve knowledge and how that knowledge is represented in an RDF-based ontology. We then proceed to discuss in Section 10.3 how the ontologies in OntoShare evolve over time based on user interaction with the system and motivate our approach to user-based creation of RDF-annotated information resources. The way in which OntoShare can help to locate expertise within an organization is then described, followed by a discussion of the sociotechnical issues of deploying such a tool. Finally, a planned evaluation exercise and avenues for further research are outlined.

10.2 Sharing and Retrieving Knowledge in OntoShare

OntoShare is an ontology-based WWW knowledge sharing environment for a community of practice that models the interests of each user in the form of a user profile. In OntoShare, user profiles are a set of topics or ontological concepts (represented by RDF classes and declared in RDF(S)) in which the user has expressed an interest. OntoShare has the capability to summarize and extract keywords from WWW pages and other sources of information shared by a user and it then shares this information with other users in the community of practice whose profiles predict interest in the information.

OntoShare is used to store, retrieve, summarize and inform other users about information considered in some sense valuable by an OntoShare user. This information may be from a number of sources: it can be a note typed by the user him/herself; it can be an intra/Internet page; or it can be copied from another application on the user's computer.

As we will see below, OntoShare also modifies a user's profile based on their usage of the system, seeking to refine the profile to better model the user's interests.

10.2.1 Sharing Knowledge in OntoShare

When a user finds information of sufficient interest to be shared with their community of practice, a 'share' request is sent to OntoShare via the Java client that forms the interface to the system. OntoShare then invites the user to supply an annotation to be stored with the information. Typically, this might be the reason the information was shared or a comment on the information and can be very useful for other users in deciding which information retrieved from the OntoShare store to access. At this point, the system will also match the content being shared against the concepts (ontological classes) in the community's ontology. Each ontological class is characterized by a set of terms (keywords and phrases) and the shared information is matched against each concept using the vector cosine ranking algorithm (Granovetter, 1982). The system then suggests to the sharer a set of concepts to which the information could be assigned. The user is then able to accept the system recommendation or to modify it by suggesting alternative or additional concepts to which the document should be assigned.

When information is shared in this way, OntoShare performs four tasks:

1. An abridgement of the information is created, to be held on the user's local OntoShare server. This summary is created using the ViewSum text summarization tool. The summarizer extracts key theme sentences from the document. It is based on the frequency of words and phrases within a document, using a technique based on lexical cohesion analysis (Ahmed and Benbrahim, 1995). Access to this locally held summary enables a user to quickly assess the content of a page from a local store before deciding whether to retrieve the (larger amount of) remote information.
2. The content of the page is analyzed and matched against every user's profile in the community of practice. As when recommending concepts to the user, the vector cosine ranking model is used: here, however, the shared information is matched against the set of terms (words and phrases) created from the union of all terms associated with the concepts to which

has user has subscribed (i.e. the concepts which make up the user profile). If the profile and document match strongly enough, OntoShare emails the user, informing him or her of the page that has been shared, by whom and any annotation added by the sharer.

3. The information is also matched against the sharer's own profile in the same way. If the profile does not match the information being shared, the system will suggest one or more concepts which strongly match the shared information that the user can then add to their profile. Thus, OntoShare has the capability to adaptively learn users' interests by observing user behaviour.

4. For each document shared, an instance of the class *Document* is created, with properties holding meta-data including keywords, an abridgement of the document, document title, user annotation, universal resource locator (URL), the sharer's name and date of storage. (The ontological structure of the OntoShare store is described in detail in the next section).

In this way, a shared and enhanced information resource is built up in the OntoShare store based on user contributions. Given that users must make a conscious decision to store information, the quality of the information in the OntoShare store is high – it is effectively pre-filtered by OntoShare users. Thus, each user leverages the assessment of the information made by all the other users.

10.2.2 Ontological Representation

We said above that each piece of shared information leads to the creation of a new entry in the OntoShare store and that this store is effectively an ontology represented in RDFS and RDF. We now set this out in more detail. RDFS is used to specify the classes in the ontology and their properties. RDF is then used to populate this ontology with instances as information is shared. Figure 10.1 shows a slightly simplified version of the ontology for a community sharing information about the Semantic Web, along with an example of a single shared document ('Document_1').

It is not our intention to describe each class and property and their function here but we will mention a few key aspects. First, notice *Concept* and its subclasses: this is the set of concepts that the community of practice at hand is interested in. Note that in the current version of OntoShare, the concept structure is limited to a strict hierarchy. Another key class is *Document*, which is the class used to represent shared information: each document shared generates an instance of *Document* with the set of properties shown. *Document_1*, for example, was stored by John Smith into the concept *RDF* with the

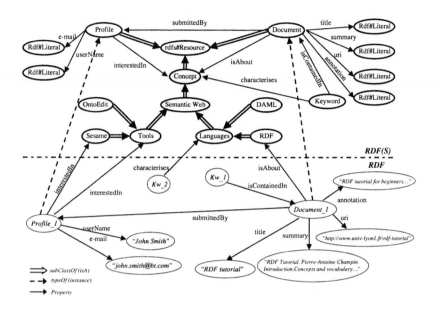

Figure 10.1 Ontological structure in OntoShare

annotation 'RDF tutorial for beginners...' with the summary and URI as shown in Figure 10.1. It also has a set of keywords associated with it. (For simplicity, note that here we show only one keyword *Kw_1*, which is an instance of the class *Keyword*, as is *Kw_2* and furthermore that the instance (typeOf) relation is not shown for these keywords, nor is the fact that Keyword is a subclass of rdfs#Resource). The third central class is *Profile*, instances of which represent user information, including the concepts in which they are interested, their names and email addresses. *Profile_1*, for example, is the profile of a user with name 'John Smith'. Finally, note that keyword Kw_2 is one of (possibly many) terms (words and phrases) which characterize the concept *Language*.

Below we include excerpts from the RDFS and RDF (in XML notation) used to represent the ontology depicted above. We see the declarations of the classes *Document, Profile* and *Keyword* in RDFS, followed by the descriptions of *Document_1* and the user profile of John Smith in RDF.

```
<?xml version="1.0" encoding="UTF-8" ?>
<rdf:RDF
  xmlns:rdf="http://www.w3.org/1999/02/22-rdf-
  syntax-ns#"
```

```
xmlns:rdfs="http://www.w3.org/2000/01/rdf-
schema#"
xmlns:ontoshare="http://www.bt.com/ontoshare#">

<!-- *************** RDFS SCHEMA *************** -->
<rdfs:Class rdf:ID="Document" />
<rdfs:Class rdf:ID="Profile" />
<rdfs:Class rdf:ID="Keyword" />

<!-- Document properties -->
<rdf:Property rdf:ID="submitted_by">
 <rdfs:domain rdf:resource="#Document" />
 <rdfs:range rdf:resource="#Profile" />
</rdf:Property>
...............
...............
<!-- *************** RDF DATA *************** -->

<!-- DOCUMENTS -->
<Document rdf:ID = "Document_1">
 <title>RDF Tutorial</title>
 <uri>http://www710.univ-lyon1.fr/~champin/rdf-
tutorial</uri>
 <submitted_by>#Profile_1</submitted_by>
 <summary>the summary of this doc goes
here</summary>
 <isAbout rdf:resource="#RDF" ontoshare:ID="7" />
 <annotation>RDF tutorial for beginners
 ...</annotation>
</Document>

<!-- PROFILES -->
<Profile rdf:ID="Profile_1">
 <user_name>John Smith</user_name>
 <email > john.smith@bt.com</email>
 <interestedIn rdf:resource=
 "#Sesame" ontoshare:ID="5" />
 <interestedIn rdf:resource=
 "#Tools" ontoshare:ID="2" />

</Profile>
...............
...............
```

10.2.3 Retrieving Explicit Knowledge in OntoShare

In this section, we discuss the ways in which OntoShare facilitates access to and the automatic sharing of the information shared by users.

10.2.3.1 Email Notification

As described above, when information is shared in OntoShare, the system checks the profiles of other users in the community of which the user is a member. If the information matches a user's profile sufficiently strongly, an email message is automatically generated and sent to the user concerned, informing the user of the discovery of the information. Thus, in cases where a user's profile indicates that they would have a strong interest in information shared, they are immediately and proactively informed about the appearance of the information.

10.2.3.2 Searching the Community Store – Accessing Information and People

Via a button on their OntoShare home page, a user can supply a query in the form of a set of keywords and phrases in the way familiar from WWW search engines. Users can also request that OntoShare only returns documents added by a particular user or added to particular concepts. A search window is shown in Figure 10.2. OntoShare then retrieves the most closely matching pages held in the OntoShare store, using a vector space matching and scoring algorithm (Granovetter, 1982).

The system then displays a ranked list of links to the pages retrieved and

Figure 10.2   Searching the OntoShare store

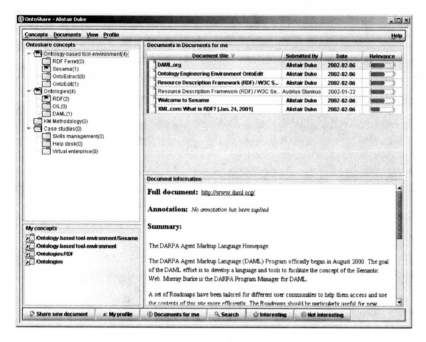

Figure 10.3 Typical OntoShare home page

their abridgements, along with the scores of each retrieved page. Any annotation made by the original sharer is also shown. Importantly, the user can elect to simultaneously search for other users by selecting the appropriate check box. We will have more to say about this capability to identify other *users* as well as *information* in Section 4 when we look at accessing *tacit* knowledge via other users using OntoShare.

10.2.3.3 Personalized Information

A user can also ask OntoShare to display 'Documents for me' as shown in the top right pane of Figure 10.3. The system then interrogates the OntoShare store and retrieves the most recently stored information. It determines which of these pages best match the user's profile. The user is then presented with a list of links to the most recently shared information, along with a summary, annotations where provided, date of storage, the sharer and an indication of how well the information matches the user's profile (the thermometer-style icon in Figure 10.3).

In addition, two buttons are provided (on the button bar at the bottom of the screen in Figure 10.3) so that the user can indicate interest or disinterest in a particular piece of information – this feedback will be used to modify the user's profile. At this point, the system will match the content of the current document against each concept (ontological class) in the community's ontology. As described above, each ontological class is characterized by a set of terms (keywords and phrases) and the shared information is matched against the term set of each concept using the vector cosine ranking algorithm (Harman, 1992). The system then identifies the set of zero or more concepts that match the information above a given ranking threshold and suggests to the sharer that this set of concepts be added to or removed from their profile in the cases of user interest or disinterest, respectively. The user is then free to accept the system recommendation or to modify it by selecting from the set of suggested concepts.

Two further operations are possible on documents presented to the user. These operations are selected from the 'Documents' menu. First, a user can add their own annotation to information stored by another user. Second, a user can request that OntoShare identifies other users with an interest in the information under consideration.

This 'Documents for me' information is in fact displayed on the user's OntoShare home page, so that whenever they access the system, they are shown the latest information. Figure 10.3 is a typical OntoShare home page.

10.3 Creating Evolving Ontologies

In Section 10.2, we described how, when a user shares some information, the system will match the content being shared against each concept (class) in the community's ontology. Recall that each ontological class is characterized by a set of terms (keywords and phrases) and that following the matching process, the system suggests to the sharer a set of concepts to which the information could be assigned. The user is then able to accept the system recommendation or to modify it by suggesting alternative concept(s) to which the document should be assigned. It is at this point that an opportunity for ontology evolution arises.

Should the user indeed override the system's recommended classification of the information being shared, the system will attempt to modify the ontology to better reflect the user's conceptualization, as follows. The system will extract the keywords and keyphrases from the information using the ViewSum system mentioned above. The set of such words and phrases are then presented to the user as candidate terms to represent the class to which the user has assigned the information. The user is free to select zero or more terms from this list and/or type in words and phrases of his own. The set of terms so identified is then added to the set of terms associated with the given concept, thus modifying its characterization.

We call this approach usage-based ontology evolution and in this way the characterization of a given concept evolves over time, this evolution being based on input from the community of users. We believe that this ability to change as users' own conceptualization of the given domain changes is a powerful feature which allows the system to better model the consensual ontology of the community.

As well as usage-based evolution, we have seen above how users also indirectly annotate the information as a side-effect of sharing it with the community and we discuss and motivate this approach below.

Pragmatically speaking, it is the case at the time of writing that only a very small proportion of WWW- and intranet-based information resources are annotated with RDF (meta)data. It is therefore beneficial to provide a system wherein such annotation effectively occurs as a side-effect of normal usage.

Another important observation is that it is in the general case impossible to cover the content of a document exhaustively by an RDF description. In practice, RDF descriptions can never replace the original document's content: any given RDF description of a set of resources will inevitably give one particular perspective on the information described. Essentially, a meta-data description can never be complete since all possible uses for or perspectives on data can never be enumerated in advance.

Our approach accommodates this observation, however, in the sense that each community will create its own set of meta-data according to its own interest in and perception of information that is added to its store. It is very possible that the same information could be shared in two separate communities and emerge with different meta-data annotations in each.

10.4 Expertise Location and Tacit Knowledge

In Section 10.2, we focused on the *technical* aspects of OntoShare and on the sharing and storing of explicit knowledge. Explicit knowledge we take to be that knowledge which has been codified in some way. This codification can take place in many different media (paper, WWW page, audio, video, and so on). In the context of OntoShare, by explicit knowledge, we mean the information shared in OntoShare, along with the meta-information associated with it such as the sharer, the annotations attached to it, and so forth. We now turn to the *social* aspects of the system and tacit knowledge.

A large amount of the knowledge within an organization may of course not be codified: it may be personal, context-specific and difficult to write down, and may be better transmitted through a master-apprentice 'learning by watching and copying' arrangement. Such knowledge is referred to as *tacit* knowledge (Polyani, 1966). When tacit knowledge is difficult to make explicit (codify), we need to find new ways of transmitting the knowledge through an organization. Failure to do so can lead to loss of expertise when people leave, failure to benefit from the experience of others, needless duplication of a learning process, and so on.

One way in which a system such as OntoShare can encourage the sharing of tacit knowledge is by using its knowledge of the users within a community of practice to put people who would benefit from sharing their (tacit) knowledge in touch with one another automatically.

One important way we gain new insights into problems is through 'weak ties', or informal contacts with other people (Granovetter, 1974, 1982). Everyone is connected to other people in social networks, made up of stronger or weaker ties. Stronger ties occur between close friends or parts of an organization where contact is maintained constantly. Weak ties are those contacts typified by a 'friend of a friend' contact, where a relationship is far more casual. Studies have shown that valuable knowledge is gathered through these weak ties, even over an anonymous medium such as electronic mail and that weak ties are crucial to the flow of knowledge through large organizations. People and projects connected to others through weak ties are more likely to succeed than those not (Constant et al., 1996; Hansen, 1997).

User profiles can be used by the OntoShare system to enable people to find other users with similar interests. The user can request OntoShare to show them a list of people with similar interests to themselves. OntoShare then compares their profile with that of every user in the store and a list of names of users whose interests closely match their own. Each name is represented as a hypertext link which when clicked initiates an email message to the named user. Recall that profiles in OntoShare are a set of phrases and thus the vector space model can be used to measure the similarity between two users. A threshold can then be used to determine which users are of sufficient similarity to be deemed to 'match'.

This notion is extended to allow a user to view a set of users who are interested in a given document. OntoShare determines which members of the community 'match' the relevant document above a predetermined threshold figure and presents back to the user a list of user names. As before, these names are presented as hypertext links, allowing the user to initiate an email message to any or all of the users who match the document. Figure 10.4 shows typical output from this process.

In addition, as already mentioned, a user can carry out a keyword search on other users and thus identify users with an interest in a particular subject.

In this way, OntoShare, while not claiming to actually capture tacit knowl-

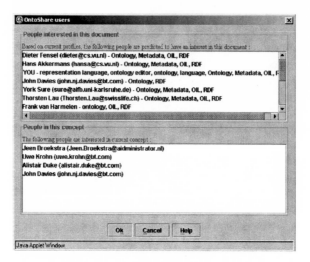

Figure 10.4   Finding users in OntoShare

edge, provides an environment which actively encourages the sharing of tacit knowledge, perhaps by people who previously would not otherwise have been aware of each other's existence.

10.5 Sociotechnical Issues

10.5.1 Tacit and Explicit Knowledge Flows

In the organizational literature, it is common to distinguish between different types of knowledge along two axes: tacit and explicit knowledge, as discussed in the previous section; and group and individual knowledge.

Clearly, knowledge management is concerned with the sharing of knowledge and hence the transfer of individual knowledge to group knowledge. This much is uncontentious but the means by which this may be achieved is less so. Nonaka (1994) is typical of much of the literature when he says 'While tacit knowledge held by individuals may lie at the heart of the knowledge creating process, realising the practical benefits of that knowledge centres on its externalization…', where by externalization is meant the conversion of tacit knowledge to explicit. Therefore, Nonaka is keen to attempt facilitation of this conversion process.

However, as we have discussed above, we view tacit knowledge as essentially of a different type from explicit knowledge. It follows from this that tacit knowledge cannot necessarily be 'externalized'. Space does not permit a full

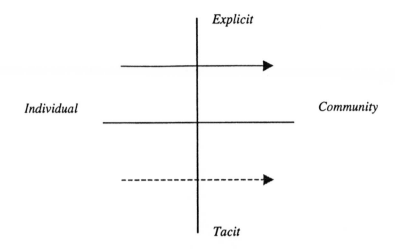

Figure 10.5   Knowledge flows in OntoShare

discussion of the epistemological issues here and the reader is referred to Cook and Seely-Brown (1998) for further details.

In OntoShare, the main focus is on explicit knowledge, which is transferred via the sharing of information (along with additional meta-information). Tacit knowledge transfer is not directly supported in OntoShare but is, however, facilitated by allowing and encouraging users to contact others with shared interests or concerns as shown in Figure 10.5.

10.5.2 Virtual Communities

There is no blueprint to guarantee a successful virtual community. In our experience, there are a number of relevant factors, some of which are inter-related. We discuss these factors briefly below.

10.5.2.1 Reward Collaborative Behaviours

This can be done in a number of ways. Collaboration in appropriate communities can be written into job descriptions and/or personal objectives and thus become part of an organization's appraisal processes. Less formal approaches could include the award of monthly prizes for the most effective contributions to a community, perhaps selected by a senior manager.

10.5.2.2 Identify a Senior Management Champion and Gain Local Management Support

We have found it almost universally true that senior management buy-in and active promotion is very helpful. A degree of care must be taken, however, to ensure that involvement from senior figures does not jeopardize the informal nature of the communities. Studies at Xerox found that informality was key to creating trust and a greater tendency to information exchange in many situations (Seely-Brown and Duguid, 1991).

10.5.2.3 Consider Job Types and Cultures

In practice, it is usually the case that many different cultures exist within the overarching culture of a large organization. One needs to understand the culture (and of course information needs) of candidate job groups in order to assess the areas where communities of practice are likely to be most beneficial. In one organization, we trialled a collaborative working solution with a community of engineers and a community of sales people: The engineers' community flourished, while that of the sales people withered on the vine. When we looked a little more closely at the nature of the communities, we uncovered the attributes summarized in Table 10.1. Of course, this summary would not necessarily apply universally: it does not even apply to all the members of the groups we were working with but nevertheless explains the success of one group and relative failure of the other.

Table 10.1   Job groups and culture

Engineers	Sales people
Job spanned multiple customers	Rarely concerned with more than one customer
Motivation tied to best quality solutions	Motivated by unique solutions and 'being first'
Schooled in total quality management and re-use of best practice	Competitive by personality (and selection) Compensation schemes reinforce competitive attitudes

10.5.2.4 Technology drivers

It has become common place in knowledge management circles that 'technology is only 20% of the solution, the rest is culture'. While the sentiment that we need a wider perspective than just technology is indisputably correct, it reveals the assumption of a dichotomy between technology and organizational culture which does not exist. Rather, technology-based tools are among the many artefacts entwined with culture, whose use both affects and is affected by the prevailing cultural environment. A holistic view is required and technology often plays a larger part in cultural factors than is sometimes acknowledged. We should never ignore the possibility of the introduction of a new technological capability driving new behaviours. An example of this would be a geographically dispersed team exhibiting a low level of collaborative behaviour: the introduction of the right tools can increase the level of collaboration significantly. A good discussion of the inter-relationship between technology and culture can be found at http://www.deepwoods.com.

10.5.2.5 Identify Facilitators

In a number of communities, we have seen great benefit in assigning 'early adopters' a facilitation role. In OntoShare, for example, this would typically involve the facilitator sharing items of the right content (and at the right technical level) on a regular basis, particularly during the early stages of the tools use.

10.5.2.6 Build trust

Trust between members of a community of practice is of course essential if people are to participate fully (indeed, it is highly doubtful whether a group of individuals comprise a genuine community without it). In some trials, members of communities will be well known to one another. Where this is not the case, it may be appropriate to bring people (physically) together to begin to build mutual trust and to set expectations for the community. It should always be made clear that no kind of censure will result from 'incorrect' participation in the community. Trust is of course a highly complex issue in its own right and the reader is referred to Maxwell (2000) for further discussion.

10.6 Evaluation and Further Work

OntoShare is a recently developed system and at the time of writing a formal evaluation has just started which we describe briefly here. The user group for the study will consist of approximately 30 researchers, developers and technical marketing professionals from the research and development arm of a large telecommunications firm. The interests of the users fall into three main groupings: conferencing, knowledge and information management and personalization technologies. It is felt that three separate yet overlapping topic areas will constitute an interesting mix of interests for the purposes of the trial.

The case study will commence with a workshop involving the practitioners in order to develop an ontology that encompasses the research fields with particular emphasis upon the overlap between them. OntoEdit (described in Chapter 7) will be used to create the ontology for the research areas. This will then be uploaded to Sesame (Chapter 5), allowing it to be viewed by the Spectacle tool (described in the previous chapter) and used as the ontology in OntoShare (which contains a module for reading ontological information from Sesame). The use of Sesame also allows access to the ontology for other ontology-based tools. The ontology will automatically evolve and extend over the course of the study as documents are added to OntoShare. The effectiveness of this evolutionary process will be considered in the evaluation exercise. Qualitative and quantitative measures of the trial are being devised. The main evaluation criterion is to what degree the application of tools and methodology can ensure that knowledge discovered by individuals can be transferred to the most appropriate members of the user group. An interesting secondary outcome we wish to look at is the extent to which the ontology built up by the community is useful to other users in other contexts. In this regard, we plan to offer a searching and browsing facility over the community's information using the QuizRDF system (see Chapter 8) for other users outside the community.

Research and development of OntoShare is ongoing. A particular area of focus currently is the ontological structure: a strict hierarchy of concepts about which the communities wants to represent and reason may prove ultimately limiting and various possibilities for allowing a more expressive concept map are under consideration. One such is that OntoShare will be developed beyond the sub-class/super-class concept hierarchy with *isRelatedTo* properties, allowing 'horizontal' links between concepts. The exploitation of this additional information is again a matter for further research. One proposal is that when seeking to match users to other users, the system can use some notion of tree-matching, taking into account the concepts in the users' profiles as well as not only the *isA* (*subClassOf*) links but also the *isRelatedTo* links. A further research area is the automatic identification and incorporation of *new* concepts as they emerge in the community. Work on this area is, however, at a very early stage.

10.7 Concluding Remarks

We have described OntoShare, an ontology-based system for sharing information among users in a virtual community of practice. We motivated the use of

Semantic Web technology for knowledge management tools and described how ontologies in OntoShare are defined in RDF and RDF(S). Communities are able to automatically share information and thereby create RDF-annotated information resources. Furthermore, these information resources are then of course available to other RDF-based tools for processing.

Importantly, the ontology used by a given community in OntoShare can evolve over time based on the concepts represented and the information that users choose to associate with particular concepts. This is a significant advantage over a community attempting to reach consensus on a set of concepts and how they relate to one another at the outset that is then difficult or impossible to change. Much remains to be done in this area, however, particularly with regard to the introduction of new concepts. In addition, users have personal profiles according to the concepts in which they have declared an interest and these profiles also evolve automatically, seeking to match more closely a user's information needs and interests based on the usage they make of the system.

We concluded by indicating some further directions of research and briefly discussed an ongoing evaluation of the system.

11

Ontology Middleware and Reasoning

Atanas Kiryakov, Kiril Simov and Damyan Ognyanov

The ontology middleware discussed in this chapter can be seen as 'administrative' software infrastructure that makes the rest of the modules in a knowledge management toolset easier to integrate into real-world applications. The central issue is to make the methodology and modules available to society as a self-sufficient platform with mature support for development, management, maintenance, and use of middle-size and large knowledge bases.[1]

This chapter starts with an explanation of the required features of ontology middleware in the context of our knowledge management architecture and the terminology used. In Section 11.2 the problem of versioning and tracking change is discussed. Section 11.3 presents the versioning model and its implementation that is developed in the project, and Section 11.4 describes the functionality of the instance reasoning module.

11.1 Ontology Middleware: Features and Architecture

The ontology middleware system serves as a flexible and extendable platform for knowledge management. It should provide infrastructure with the following features:

- A *repository* providing at least the basic storage services in a scalable and reliable fashion.
- Support for *plugable reasoning modules* suitable for various domains and applications. This ensures that within a single enterprise or computing environment one and the same system may be used for various purposes

---

[1] See the discussion on ontology vs. instance data vs. knowledge base below.

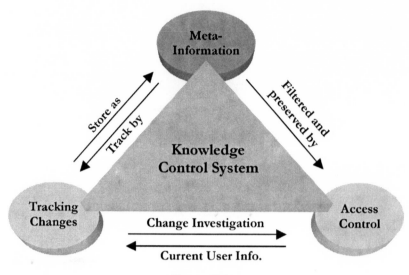

Figure 11.1

so providing easy integration, interoperability between applications, knowledge maintenance and re-use.

- *Multi-protocol client access* to allow different users and applications to use the system via the most efficient 'transportation' media.
- *Knowledge control.* The following features are considered:

    - versioning (tracking changes) in knowledge bases;
    - access control (security) system;
    - meta-information for knowledge bases.

- These features are interrelated to each other as depicted in Figure 11.1.

We call the composition of these three functions a knowledge control system (KCS). It provides knowledge engineers with the same level of control and manageability of the knowledge in the process of its development and maintenance as source control systems (such as Concurrent Versions System) provide for software. From the perspective of the end-user applications, KCS can also be seen as equivalent to database security, change tracking[2] and auditing systems. A KCS should be carefully designed to support these two distinct use cases.

---

[2] The change tracking is often called 'cataloguing'.

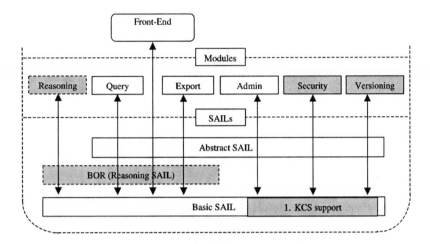

Figure 11.2

11.1.1 Place in the On-To-Knowledge Architecture

The ontology middleware module (OMM) is implemented as an extension of the Sesame architecture that fulfils the repository role and provides a good basis for multi-protocol access and plugable storage and inference layers. The exact approach undertaken can be seen in Figure 11.2; it represents a modification of the bottom part of the Sesame architecture diagram presented in Figure 5.3.

The grey boxes denote the additional modules and extensions developed. The arrows depict how the requests of the different modules are passing through the different SAIL interfaces. The boxes with dashed borders (Reasoning and BOR – the DAML+OIL reasoner) are optional, that is, not an obligatory part tightly coupled with the Sesame architecture. Those are only necessary for DAML+OIL reasoning.

The interfaces of the OMM are designed to fit in the stacked SAIL architecture by extending it with their specific functionality. Some of the Sesame functional modules are modified so to benefit directly from the interfaces in the OMM SAIL extension. The implementation of BOR is an additional SAIL. Its full potential is accessible via a specific functional interface (described in Section 11.4). On the other hand, the BOR supports the SAIL interface so the rest of the modules can also interoperate with it in the same way as they do with standard RDF(S) supporting SAIL. For instance, the Query module can perform queries against DAML+OIL repositories without changes to its interfaces.

The above architecture allows a transparent manipulation of the repository for the existing tools. Each application can work with the repository through the standard Sesame modules – the middleware remains transparent in this case. In addition, the repository behaviour can be changed if the application is working directly with the new reasoning, versioning, and security modules or gaining access to the underlying low-level programming interfaces.

11.1.2 Terminology

A number of justifications in terminology are necessary. An almost trivial but very important question is 'What do the knowledge management tools support: ontologies, data, knowledge, or knowledge bases?' A simple and correct answer is 'All of these.' Here we provide a simplified interpretation of these terms.

Ontology is the basic knowledge formally defining the model (schema, conceptualization) relevant for a certain knowledge domain. Such a model allows the information relevant to a more or less general set of applications considering this domain to be represented as much as possible in an adequate, effective, structured, and non-redundant way. The ontologies are usually defined in a kind of formal logical or schema-definition language and at least define the types of entities to be considered together with some characteristics, constraints, and relations applicable to them. An ontology may or may not include: type hierarchies, attribute definitions and restrictions, induction or other rules. Examples of ontologies could be a database schema or a definition of a product catalogue structure and categories.

Data or *instance data* is the particular information to be managed – it usually considers specific situations or phenomena. The data should always be structured according to the ontology (if this is defined, of course) and comply with its restrictions. In the ideal case, a big number of different data sets and applications use one and the same ontology. While the instance data and the applications may regularly change to reflect changes in the situation or particular needs, the ontologies are expected to be much more stable so as to provide a coherent interpretation of the data and easy integration among the applications.

Knowledge is a hard to define philosophical category that is often interpreted in different ways according to the context and purpose of use. It is the case that both ontologies and data can be considered as knowledge, so, it is often used as a generic term denoting both types of information. *Knowledge base* is a term used in a way similar to database, so, not surprisingly it may denote both a specific body of knowledge as well as the software system for its management. In contrast to databases, knowledge base management systems are expected to support some inference mechanisms in order to be able to provide explicit facts that logically follow from the existing knowledge but not formally asserted.

Repository is also often used as a synonym of database or knowledge base or just as a more generic term that denotes any of them, as well, as hybrid and marginal systems. So, everything discussed in the previous paragraph is also applicable for repositories.

For the purpose of compliance with the terminology used in the Sesame RDF(S) repository, the term *repository* is used here to denote a compact body of knowledge that could be used, manipulated, and referred to as a whole. Such may contain (or host) both ontological statements and instance data.

In the knowledge management community there is a rich diversity of almost equivalent terms used to reference similar modelling primitives – this is result of the diversity paradigms with their own vocabularies. We use mostly the RDF(S) terminology and when necessary DAML+OIL. So, here we use: *Class* for any concepts, types and other unary predicates; *Property* for any binary relations, attributes, slots, etc. More detailed discussion on the terminology diversity can be found in Kiryakov et al. (2001).

11.2 Tracking Changes, Versioning and Meta-information

> The Big Brother is watching you!
> 1984, George Orwell

The problem of tracking changes within a knowledge base is addressed in this section. It is important to clarify that higher level evaluation or classification of updates (considering, for instance, different sorts of compatibility between two states or between a new ontology and old instance data) is beyond the scope of this work (see Ding et al., 2001, Section 2.2). Tracking changes in the knowledge (as discussed here) provides the necessary basis for further analysis. As an example, in order to judge the compatibility between two states of an ontology, a system should be able to at least retrieve the two states and/or the differences between them.

An overview of the related work is followed by formal statement of the requirements. Next, a model that satisfies these requirements is presented, complemented with comments on the implementation approach. In summary, the approach taken can be briefly characterized as 'versioning of RDF on a structural level in the spirit of the software source control systems".

11.2.1 Related Work

Here we briefly review similar work, namely, several other studies related to the management of different versions of complex objects. In general, although some of the sources discuss closely related problems and solutions, there is no one addressing ontology evolution and version management in a fashion allowing granularity down to the level of specific resources and statements (or similar constructs) and also able to capture interactive changes in knowledge repositories such as asserting or retracting statements.

One of the studies that provides a methodological framework close to the one needed here is Kitcharoensakkul and Wuwongse (2001). The authors model a framework, which is designed to handle the identification, control, recording, and tracking of the evolution of software products, objects, structures, and their interrelationships. The paper investigates the different models and versioning strategies for large scale software projects and presents a way to express the meta-information and the impact of a single change over the various components of the project in RDF(S) – in this case used just for representation of the related meta-information; the objects being tracked are pieces of software.

Database schema evolution and the tasks related to keeping schema and data consistent to each other can be recognized as very similar to ours. A detailed and formal study on this problem can be found in Franconi et al. (2000a); it presents an approach allowing the different sorts of modifications of the schema to be expressed within suitable description logic. More detailed information about the reasoning and other related tasks can be found in Franconi et al. (2000b). Similar issues are discussed also in Benatallah and Tari (1998).

Unsurprisingly, some of the most relevant studies were done under the On-To-Knowledge project; among the reports concerning various aspects of the knowledge management, the most relevant is Ding et al. (2001), mentioned earlier in this section.

11.2.2 Requirements

The top-level requirements towards a versioning model for knowledge management in the contexts targeted by the On-To-Knowledge project are enumerated below:

1. To allow tracking of changes in both ontologies and instance data with fine granularity.
2. To allow work with, revert to, extraction of, and branching of different states and versions of a knowledge base. This to be possible without any loss or corruption of information related to other states or versions.

3. To be independent from the serialization of the knowledge. In other words a change of the ordering or formatting of the knowledge that does not cause changes in its formal interpretation not to be considered as a change of the state. This requirement should not be understood in the sense of a necessity for continuous evaluations of logical equivalence, but rather as a need for abstraction from re-ordering of statements or terms and other similar cosmetic changes which are syntactically detectable as sense-preserving in a declarative representation.

4. To be applicable for various platforms or medias. The mechanism to allow versioning of repositories (where statements can be asserted and retracted in an interactive way), but to be also adaptable for use cases when the different states of a repository (a body of knowledge) are represented as, say, XML files.

5. To be applicable for knowledge represented in RDF(S), DAML+OIL, and similar ontology languages.

6. To be as much as possible independent from the semantics of a specific state or variant of a single language. Taking into account the current state of the development of the Semantic Web vision and the related languages, specifications, and tools, development of a scheme particularly tuned for language such as DAML+OIL seems not to be feasible.

7. To be simple and intuitive, optionally transparent, to allow use of the knowledge in a manner abstracting from the change tracking. Thus, not to unnecessarily increase the complexity of simple applications.

8. To allow versioning of the meta-information as well. (Motivation provided below.)

9. To allow naming, addressing, and keeping meta-information for specific states. Well-structured meta-information considering various characteristics of a certain state (or version) of a knowledge base to be supported in a flexible way.

10. To cause minimal overhead to the volume of the data.

11. To provide a basis for further evaluations of the changes, such as checking the consistency of a state and various types of compatibility between states of ontologies and data (as in Ding et al., 2001).

11.3 Versioning Model for RDF(S) Repositories

A versioning model for RDF(S) repositories is presented. To make it more explicit (i) the knowledge representation paradigm supported is RDF(S) and (ii) repositories are being tracked independently from the fact of whether they contain ontologies, instance data, or both. The decision to support tracking of changes, versioning, and meta-information for RDF(S) repositories has a number of consequences and requires more decisions to be taken. The most important principles are presented in the next paragraphs.

Principle 1.   *The RDF statement is the smallest directly manageable piece of knowledge.*

Each repository, formally speaking, is a set of RDF statements (i.e. triples); these are the smallest separately manageable pieces of knowledge. Arguments can be found to support that the resources and the literals are the smallest entities; it is true in a way, however, they cannot be manipulated independently. None of them can independently 'live' in a repository because they always appear as a part of a triple and never independently. The moment when a resource is added to the repository may only be defined indirectly as the same as 'the moment when the first triple including the resource was added'. Analogously, a resource may be considered as to be removed from a repository when the last statement including it gets out. To summarize, there is no way to add, remove, or update (the description of) a resource without also changing some statements while the opposite does not hold. So, the resources and the literals from a representational and structural point of view are dependent from the statements.

Principle 2.   *An RDF statement may not change – it can only appear and disappear.*

As far as the statements are nothing more than triples, changing one of the constituents, just converts it into another triple. This is because there is nothing else but the constituents to determine the identity of the triple, which is an abstract entity that is fully defined by them. Let us take for instance the statement ST1=<A, PR1, B> and suppose B is a resource, that is, a URI. Then ST1 is nothing more than a triple of the URIs of A, PR1, and B – if one of those get changed it will be already pointing to a different resource that may or may not have something in common with the first one. For example, if the URI of A was http://x.y.z/o1#A and it got changed to http://x.y.z/o1#C then the statement ST2=<C,PR1,B> will be a completely different statement.

Further, if the resource pointed by a URI gets changed, two cases can be distinguished:

- The resource changes but its meta-description in RDF does not. Such changes are outside the scope of the problem for tracking changes in formally represented knowledge.

- The description of the resource changes – it can happen if and only if a statement including this resource get changed, that is, added or removed. In such a case, there is another statement affected, but the one that just bears the URI of the same resource does not.

There could be an argument that when the object of a triple is a literal and it gets changed, this is still the same triple. However, if there is, for instance, a statement $<A,R,"abc">$ and it changes to $<A,R,"cba">$, the graph representation shows that it is just a different arc because the new literal is a new node and there could be other statements (say, $<B,P,"abc">$) still connected to the old one.

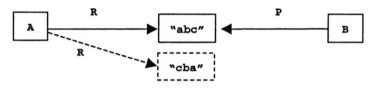

The next principle is a consequence of this.

Principle 3.   *The two basic types of updates in a repository are addition and removal of a statement.*

In other words, those are the events that necessarily have to be tracked by a tracking system. It is obvious that more event types such as replacement or simultaneous addition of a number of statements may also be considered as relevant for an RDF(S) repository change tracking system. However, those can all be seen as composite events that can be modelled via sequences of additions and removals. As far as there is no doubt that the solution proposed allows tracking of composite events (say, via post-processing of the sequence of the simple ones), we are not going to enumerate or specify them here.

Principle 4.   *Each update turns the repository into a new state.*

Formally, a *state* of the repository is determined by the set of statements that are explicitly asserted. As far as each update is changing the set of statements, it is also turning the repository into another state. A tracking system should be able to address and manage all the states of a repository.

Some notes and definitions that complement the principles stated above are presented in the following sub-sections.

11.3.1 History, Passing through Equivalent States

The history of changes in the repository can be defined as a sequence of states, as well as a sequence of updates, because there is always an update that turned the repository from one state to the next. It has to be mentioned that in the history, there can be a number of equivalent states. It is just a question of perspective whether we consider those as one and the same state or as equivalent ones. Both perspectives have some advantages for some applications. We accepted that there could be equivalent states in the history of a repository, but they are still managed as distinguishable entities. Although it is hard to provide formal justification for this decision, the following arguments can be presented:

- For most applications, it is not typical for a repository to pass through equivalent states often. Although possible, accounting for this phenomenon is not obviously worth taking into account as finding equivalent states could be a computationally very heavy task.
- If necessary, equivalent states can be identified and matched or combined via post-processing of a history of a repository.

11.3.2 Versions are Labelled States of the Repository

Some of the states of the repository could be *versions*, which could be any state, without any formal criteria and requirements – it completely depends on the user's or application's needs and desires. Once defined as a version, the state becomes a first class entity for which additional knowledge can be supported as a meta-information (as described below).

11.3.3 Implementation Approach

Let us now present the schema for tracking changes in a repository. For each repository, there is an *update counter* (UC), an integer variable that increases its value each time the repository get updated, which in the basic case means when a statement get added to or deleted from the repository. Let us call each separate value of the UC an *update identifier* (UID). Then for each statement in the repository, when the UID was added and removed will be known – these two values determine the 'lifetime' of the statement. It is also the case that each state of the repository is identified by the corresponding UID.

If the UIDs that determine the 'lifetime' of each statement are kept, for each state it will be straightforward to find the set of statements that determine it – those that were 'alive' at the UID of the state being examined. As far as

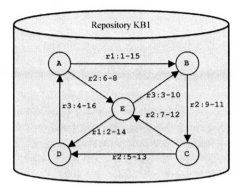

Figure 11.3

versions are nothing more than labelled states, for each one there will also be a UID that uniquely determines the version.

This approach can be demonstrated with the sample repository KB1 and its "history". The repository is represented as a graph in Figure 11.3; the lifetime of the statements is given separated by semicolons after the property names. The history is presented as a sequence of events in the format UID:nn {add|remove} <subj, pred, obj>. Two 'snapshots' of states of the repository for UIDs 2 and 8 are shown in Figure 11.4.

It is an interesting question as to how do we handle in the above model, multiple additions and removals of one and the same statement, which in a sense periodically appears and disappears form the repository. We have undertaken the approach to consider them as separate statements, because of reasons

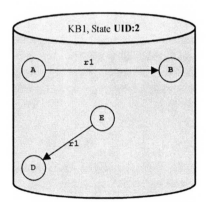

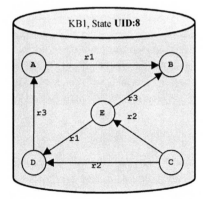

Figure 11.4

similar to those presented for the support of distinguishable equivalent statements.

11.3.3.1 Batch Updates

We call a batch update the possibility of stopping the update counter of the repository, so as not to increment its value for a number of consequent updates. This feature can be very important for cases when it does not make sense that individual updates are tracked one by one. An example could be assertion of a DAML+OIL element that is represented via a set of RDF statements none of which can be interpreted separately (see Section 11.4).

Another example for a reasonable batch update would be an application that works with the repository in a transactional fashion – a series of updates are bundled together, because according to the logic of the application, they are closely related. Finally, batch updates can also be used for file imports.

11.3.3.2 Versioning and Meta-information for Inferred Statements

There are cases when the addition of a single statement in the repository leads to the appearance of several additional statements in it. An example is the addition of the statement `ST1=<B, rdfs:subClassOf, C>` which leads to the addition of two new statements `ST2=<B, rdf:type, rdfs:Class>` and `ST3=<C, rdf:type, rdfs:Class>`.

New statements can also appear when an external ontology is imported into the repository either by a `xmlns:prefix="uri"` attribute of an XML tag in the XML serialized form of the ontology or by a `daml:imports` statement found in the header of a DAML+OIL ontology. In each of those cases, the inferred or imported statements in the repository should be treated as read-only and thus the users of the repository cannot change them. Also all these statements appear and disappear in the repository at the same moment as the statement that causes their inference or import. An additional note about the imported statements relating to the security: these statements should be recognized as external, and not belonging to the repository and thus we can avoid the application of the security policies to them.

11.3.4 Meta-information

Meta-information should be supported for the following classes: resources, literals, statements and versions. As far as DAML+OIL ontologies are also formally encoded as resources (of type daml:Ontology) meta-information can also be attached to them.

The meta-information is also modelled in RDF – something completely admissible taking into account the unrestricted meta-modelling approach in RDF(S). A number of objections against such an approach can be given. First, it raises the number of meta-layers and so makes the representation more abstract and hard to understand. However, adding meta-information always requires one more layer in the representation, so, making it via extensions of the same primitives used for the 'real data' (instead of defining some new formalisation) can even be considered as a simplification. Second, it makes confusion possible and may introduce technical difficulties, say, because of intensive use of heavy expressive means such as reification.

The schema proposed below handles these problems to some degree and provides some advantages:

- It is probably the most typical role of RDF to be used for encoding of meta-information.
- One and the same technology can be used for viewing, editing and management of both knowledge and meta-information. Any RDF(S) reasoners and editors can handle the meta-information without special support for it.
- Queries including both knowledge and meta-information are straightforward. So, lookup of knowledge according to conditions involving both meta-information and 'real' knowledge is possible. Imagine a situation when a complex ontology is being developed and there is meta-information supporting this process, say, a meta-property 'Status' (with possible values 'New', 'Consistent with the ontology', 'Verified against the sample data', 'Done') being attached to each class. Then a lookup of all classes that are subclasses of C and have status 'New' will be just a typical query against the RDF repository.
- Representing the meta-information as RDF could be done in a flexible way that allows it to be customized for the specific needs of the use case.

11.3.4.1 Tracking Changes in the Meta-information

An important decision to be taken is whether changes in the meta-information should be tracked. The resolution proposed here is as follows. Changes in the meta-information should be considered as regular changes of the repository, so, to turn it from one state to another. Here are few arguments backing this position:

- There are a number of cases when the only result of serious work on an ontology is just a single change in the meta-information. Let us use again the example with the 'Status' meta-property for classes (described above.) The result of a complex analysis of the coherence of a class definition may result just in changing the status from 'New' to one of the other values. In this case, although there is no formal change in the 'real' data, something important get changed. From an ontology management point of view, it is extremely important that tracking of such changes is possible.

- It is possible that appropriate analysis can be made on demand so that changes that affect only the meta-information can be ignored. This way both behaviours can be achieved. In the case of the opposite decision (not to track changes in meta-information), no kind of analysis can reconstruct the missing information.

- An analogy with software source control systems may also provide additional intuition about this issue. If we consider the comments in the software code as meta-information, it becomes clear that the source control systems definitely consider changes in the meta-information as equal to 'real' changes in the code.

11.4 Instance Reasoning for DAML+OIL

The represented knowledge allows a (potentially) infinite number of possible uses, but within a practical system, mainly the typical usage cases are implemented efficiently. Two typical usage cases are *Ontology development* and *Ontology use*. *Ontology development* requires the following basic tasks to be supported:

- checking whether a class definition is consistent by itself or with respect to a set of other class descriptions;
- checking whether a given class definition is more general than another class definition;
- construction of an explicit hierarchy of class names on the base of their class definitions.

We call these reasoning tasks *terminological reasoning*.

Ontology use involves an already developed ontology in which the classes are defined and (possibly) the relations between them are explicitly represented (after some terminological reasoning) and instance data of much higher magnitude, say, thousands or millions of instances. This usage case requires task such as:

- to find the most specific classes that describe a partially specified instance;
- to find all instances in the data set which are instances of a given class definition;
- more complex queries trying to retrieve answers to questions involving instance data; for example, getting all pairs of instances related in some way;
- checking the consistency of the instance data with respect to the ontology.

We call these reasoning tasks *instance reasoning*.

The infrastructure initially developed under the On-To-Knowledge project separates the ontology and data representation from the reasoning over them. The storage, management and querying of ontologies and instances is handled by the Sesame system (see Chapter 5). If more expressive reasoning is necessary (in a more expressive language like OIL or DAML+OIL), then the corresponding information should be sent to an external reasoner (say, the FaCT system) that processes it and returns the answer. Although such a solution is appealing in terms of re-use of existing tools and compliance, in our view it can hardly provide a good level of performance and interoperability in cases when the ontology and/or the instance data is huge. There are two possible approaches for implementation of such an architecture:

1. Only the relevant parts of the ontology and the instance data are sent to the external reasoner. Such a solution minimizes the exchange overhead between the two system, but imposes the question how these relevant parts are determined. In general, the problem of fragmentation of an ontology and/or instance data into non-interacting chunks can require a considerable amount of reasoning.
2. The external reasoner to support its own copy of the ontology and/or instance data. In this case the reasoner duplicates a lot of the functions of the repository, one way or another everything 'known' to the repository should be passed to the reasoner.

Both approaches are in contradiction with the expectation for Sesame usage – thousands of classes and millions of instances. Thus, integration with the external reasoner seems feasible only for small ontologies.

Using only Sesame is also not a satisfactory solution because it supports only RDF(S) which has expressive power that is insufficient to support ontologies and instance data for domains and applications requiring the full inventory of ontological languages such as DAML+OIL.

These arguments motivated our position that the reasoning services has to satisfy the following requirements:

- to be aware of the semantics of DAML+OIL;
- to be efficient in the typical usage cases;
    - ontology development – terminological reasoning, usually no instances are involved,
    - ontology use – instance reasoning, stable ontology with huge instance data;
- to be in close integration with the RDF(S) repository.

In the following, we discuss the inference services necessary to support the typical uses. First, the instance reasoning services are discussed. Next, the inference services are specified in terms of high-level functional interfaces.

11.4.1 Inference Services

In this section we describe the inference services that we consider useful for instance reasoning. In the following, we use the term 'Repository' to point to the Sesame RDF(S) repository which contains a well-defined DAML+OIL repository. The inference(s) will be concerned with the ontology and the data set parts of such a repository, called instance resources.

- *Realization.* Find in an ontology the most specific classes that describe a given instance resource.
- *Instance checking.* Check whether a given instance resource is an instance of a given class.
- *Retrieval.* Find the set of instance names that are described by a given class. Besides the individuals, one can require retrieval of their components as well. A component of an instance consists of all instances connected to the given instance by some DAML+OIL property.
- *Model checking.* Check whether a given set of instances is a correct model of a given ontology. This inference service is very useful for compatibility checks between versions of ontologies and data sets.
- *Minimal sub-ontology extraction.* Find the minimal sub-ontology for which a given set of instance statements is a correct model. A minimal ontology is defined as a minimal sub-taxonomy, but also such ontology will need to include some non-hierarchical knowledge (because of generalized concept inclusion axioms). This inference service can be very useful for determining the scope of an ontology exchange, for example, when certain information (typically a set of instances) has to be exchanged between two systems (or databases or knowledge bases).

In order to make the reasoning task feasible within the time constraints, we offer one graduated approach to the tasks that we will implement. First, we classify the possible data sets with respect to their complexity. At one end of the scale, we have the data sets in which instance statements of an arbitrary complex form are presented. At the other end of the scale, we have data sets which contain only ground instance statements. Such data sets are called ground data sets.

General instance statements

    a : C   class statement: C is an arbitrary complex class expression
    (a,b) : R   property statement: R is a property name

a = b	equality statement
a ≠ b	inequality statement
∀x.C(x)	universal statement: C is an arbitrary complex class expression, a, b are instance names, x is an instance variable

The class and the universal statements provide high expressivity within the data sets. They allow one to state additional ontological restrictions over a particular data set. Usually these statements impose great computational problems to the inference procedure. Here are examples of such statements, a: (Mother ∩ (·3 has-child)) states that a is a mother with more than two children. ∀x.(¬ Mother ∪ (·3 has-child))(x) states that each mother in the current domain has more than two children.

Ground instance statements

a: A	concept statement: A is a class name
(a,b): R	property statement: R is a property name
a = b	equality statement
a ≠ b	inequality statement

Our expectations are that data sets containing only ground instance statements will allow very efficient implementation of inference procedures. Also ground data sets are typical of the case studies described in this book and other knowledge management applications. It is important to mention that universal statements are consequences from the ontological knowledge and thus they can be expected to be always presented as constraints over the instance data.

11.4.2 Functional Interfaces to a DAML+OIL Reasoner

In this section we propose a collection of interface functions necessary for effective use and maintenance of repositories of DAML+OIL represented knowledge that may contain both ontologies and data sets. Here we consider the repositories as abstract objects and the corresponding interfaces are operations working on these abstract objects and producing new abstract objects. The interfaces are defined as functions over repositories and elements. All interfaces are defined with respect to a current repository, but if it is necessary to make it explicit, an optional argument *Repository* can be used.

- *Tell interfaces.* Tell interfaces make it possible to add knowledge to the repository. Some tell interfaces do not add new information, but instead only initiate some processing over the repository.
- *Delete interfaces.* Delete interfaces allow deletion of elements from a repository. We assume that what is deleted is a DAML+OIL element

and its interpretation is unique within the repository. Only elements that are explicitly added to the repository by the user can be deleted.

- *Ask interfaces.* Ask interfaces query the repository. The ask interfaces follow the ideas of Bechhofer et al. (1999), extending their interfaces with appropriate interfaces for instance reasoning.

12

Ontology-based Knowledge Management at Work: The Swiss Life Case Studies

Ulrich Reimer, Peter Brockhausen, Thorsten Lau
and Jacqueline R. Reich

12.1 Introduction

This chapter describes two case studies conducted by the Swiss Life insurance group with the objective of proving the practical applicability and superiority of ontology-based knowledge management over classical approaches based on text retrieval technologies.

The first case study in the domain of skills management uses manually constructed ontologies about skills, job functions and education. The purpose of the system is to give support for finding employees with certain skills. The ontologies are used to ensure that the user description of skills and the machine-held index of skills and people use the same vocabulary. The use of a shared vocabulary increases the performance of such a system significantly.

The second case study aims at improving content-oriented access to passages of a 1000 page document about the International Accounting Standard on the corporate intranet. To this end, an ontology was automatically extracted from the document. It can be used to reformulate queries that turned out not to deliver the intended results. Since the ontology was automatically built, it is of a rather simple structure, consisting of weighted semantic associations between the relevant concepts in the document. We therefore call it a 'lightweight ontology'.

The two case studies cover quite different aspects of using ontologies in knowledge management applications. Whereas in the second case study an ontology was automatically derived from a search space to improve information retrieval, in the first skills management case study the ontology itself introduces a structured search space. In one case study we gathered experience in building an ontology manually, while the challenge of the other case study was automatic ontology creation. A number of the novel Semantic Web-based tools described elsewhere in this book were used to build the two systems and both case studies described have led to projects to deploy live systems within Swiss Life.

12.2 Skills Management

12.2.1 What is Skills Management?

The tacit knowledge, personal competencies and skills of its employees are the most important resources of a company for solving knowledge-intensive tasks, such as decision-making, strategic planning, or creative design. They are the real substance of the company's success (Taubner and Brössler, 2000). Therefore, establishing an electronically accessible repository of people's capabilities, experiences and key knowledge areas is one of the major building blocks in setting up enterprise knowledge management. Such a skills repository forms the basis for a skills management system (Ackerman et al., 1999) (see Figure 12.1) that can be used to:

- enable a search for people with specific skills;
- expose skill gaps and competency levels;
- direct training, education and learning opportunities as part of the career planning process;
- document that part of a company's intellectual capital that results from the knowledge of its employees.

By making employees' experiences, knowledge and skills explicit, it is easier to find out what people know or to direct people to others who can be of help. This sharing of information improves organizational productivity as well as individual performance.

Skills management is a prerequisite for supporting project staffing, and it enables the planning of professional development (Auer, 2000; Sure et al., 2000) – or, as Younker phrased it,

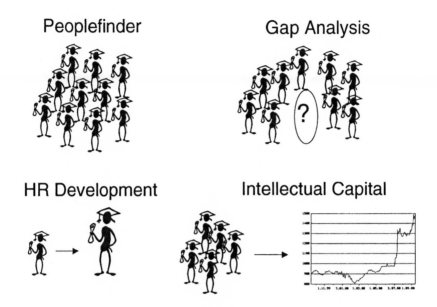

Figure 12.1   Tasks of a skills management system

> Skills management is a robust and systematic approach to forecasting, identify-
> ing, classifying, evaluating and analysing the work force skills, competencies
> and gaps that enterprises face. (Younker, 1998)

Introducing a skills management system requires effort in three dimensions.
The *technical dimension* deals with providing the necessary functionality. The
content dimension encompasses the setting up of organizational and automatic
processes for providing and updating the system content. Otherwise, outdated
experience profiles will lead to a decreased utility of the system and can
ultimately end up with a system being not used at all. The concern of the
cultural dimension is to ensure a climate of trust and openness so that employees are motivated to make their skills known – to their own and to the company's benefit.

Skills management also presupposes a critical mass of users to justify
investment in the system. A successful system may offer the means to affect
a cultural change and instill real change into the organizational mind-set and
value-set (Deiters et al., 2000; Liao et al., 1999).

A skills management system should be based on specific ontologies to be
used for entries in the skills, job function and education areas. This ensures
that terms for describing skills and for searching them are used consistently.
Other types of information, such as personal experiences, are very hard to

structure using an ontology because they are very subjective and personal. Thus, for them free text fields are required.

12.2.2 SkiM: Skills Management at Swiss Life

At Swiss Life, we developed an ontology-based skills management system (SkiM) that allows each employee to describe her or his skills, qualifications, experiences and special interests. From that description, a personal home page on the intranet is generated. A SkiM query interface combines an ontology-based search with a classical free text (keyword) search on those home pages.

In its first version, SkiM aims at finding people with a specific skills profile, and can either be used for staffing new projects, or for identifying experts who might be of help to solve a certain problem. The first prototype was used and evaluated by more than 100 pilot users from different departments. The current system is now being used by more than 150 users although the system is not yet officially introduced.

Employees are provided with structured templates to enable them to design their personal home page in a standardized format. Existing employee data from organizational databases is automatically inserted into the templates. An employee then specifies her or his skills by selecting concepts from a skills ontology and by indicating a level for each selected skill according to a scale of four levels: basic knowledge, practical experience, profound knowledge and top specialist. According to the experience made in other companies (like Credit Suisse, ABB and IBM), the approach to have employees describe their skills themselves proved to produce highly reliable information in combination with the four-level skills grade. The social pressure works as a corrective, motivating employees to be honest in describing their skills.

Besides the skills, additional details can be given, such as education, former affiliations, special interests, or projects participated in. Finally, from all given statements, a personal home page is generated on the intranet, which can then be searched. A home page currently contains (Figure 12.2):

- general information, such as name, functional unit, room, phone number;
- ontology-based details of personal skills, job function, and education;
- free text descriptions, such as qualification, current tasks, projects involved in, kind of work interested in;
- a more personal section giving, for example, information about hobbies.

Since a skills ontology can never be complete, the users of the system can make suggestions for extending the ontology as needed. Therefore, SkiM offers a dialogue, where an employee can suggest new ontology concepts as well as the place in the hierarchy where they should go. The skills ontology

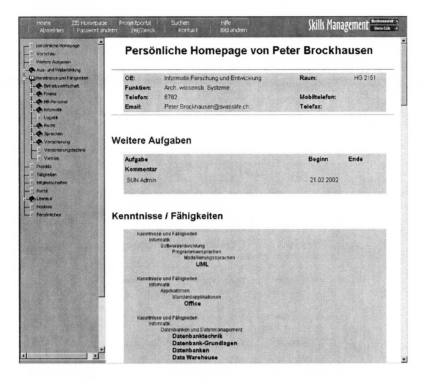

Figure 12.2 A personal home page in SkiM (top part)

administrator collects the suggestions and ensures a regular consolidation of the ontology.

Participation in SkiM is absolutely voluntary. Instead of making it obligatory we rely on the employees' motivation to become more visible within the company and thus to increase their career opportunities. In this sense, SkiM can be seen (among other things) as providing an internal job fair.

An approach complementary to ours is to identify people with certain skills by doing text mining on the documents in the intranet (Becerra-Fernandez, 2000; McDonald and Ackerman, 1998). By applying information extraction technology, the Expert Finder tool identifies and correlates names and topic areas in documents (Mattox et al., 1999). XperNet in contrast uses clustering and network analysis techniques (Maybury et al., 2000). Groups of people with related skills and interests are identified by processing the project information, publications and web pages. Expertise indicators like explicit references or citations are used in combination with counter indicators like being a member of the administrative staff.

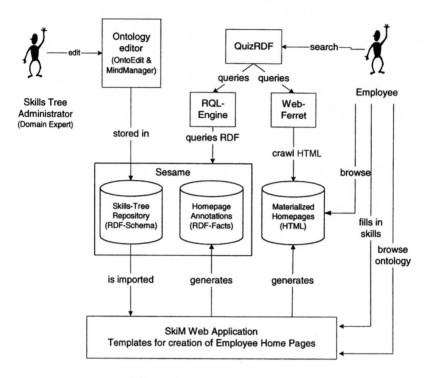

Figure 12.3   Architecture of SkiM

12.2.3 Architecture of SkiM

The SkiM system comprises several components (see Figure 12.3). The ontology editor *OntoEdit* allows an administrator to create and edit the required ontologies for skills, education and job functions. OntoEdit enables inspecting, browsing, codifying and modifying ontologies and therefore supports the ontology development and maintenance tasks. The ontologies are modelled at a conceptual level and independently of the formalism of the final representation language. OntoEdit offers views on conceptual structures, such as concepts, concept hierarchy, relations, or axioms (see Chapter 7). For the early phases of ontology development, the *MindManager* tool was used to edit the ontologies because it better supports brainstorming processes.

The web application part of SkiM allows employees to build their personalized intranet home pages by filling in the information categories given by templates. *Sesame* is an RDF/RDF Schema storage and retrieval system. Within SkiM Sesame stores the skills ontology as an RDF Schema and the instances of the ontology concepts, namely the association of skills to employees, as RDF facts. It also stores any additional RDF annotations of the home pages which serve to characterize the content of the free text fields. Sesame supports expressive querying of RDF Schema and RDF facts by means of a query engine for the RQL query language (see Chapter 6 for further details on the Sesame system).

The query interface employs *QuizRDF* (see Chapter 8) to do a combined ontology-based and free text retrieval (see Section 2.5). QuizRDF combines full text searching with querying RDF facts, in our case the skills data for each employee stored in Sesame as well as the additional annotations. Full text searching is provided to offer high recall and coverage of unannotated information, while precise ontological queries result in a high precision. Of course, a combination of both query modes is possible.

12.2.4 SkiM as an Ontology-based Approach

12.2.4.1 The Underlying Ontology

Within SkiM, three ontologies are defined: for skills, education and job function. At the moment, these ontologies are simple taxonomies (i.e. each class has only one parent) but it is intended to extend this in order to allow more advanced functionality to be added to SkiM (see Section 12.2.6). SkiM forces every skill, education or job description to be formulated by terms selected from the corresponding ontology. We thus make sure that the terms used for describing skills, education or jobs will match with query terms when SkiM users search for information. This will guarantee a high recall and precision of the result sets. Moreover, the application of ontologies is a prerequisite for comparing skills descriptions, for generating a classification of the organization's knowledge, and for doing a so-called gap analysis which identifies skills not sufficiently present in the organization but needed.

The skills ontology consists of three rather independent branches which correspond to the three organizational units that were selected for the pilot phase, that is, IT, Private Insurance and HR (see Figure 12.4). The ontologies for education and job function are not subdivided into sub-domains. Currently, the skills ontology consists of 700 concepts, the education ontology consists of 180 concepts, and the job function ontology comprises 130 concepts.

The concept hierarchies are only that part of the underlying ontology which a SkiM user sees. The complete ontology additionally includes concepts and attributes to allow the connection between employees and their descriptions. An OIL fragment that gives an impression of the whole ontology is shown in Figure 12.5.

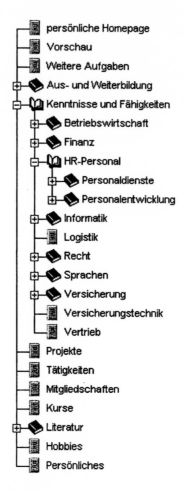

Figure 12.4   Fragment of the Swiss Life skills ontology

12.2.4.2 Ontology Building

The development and maintenance of appropriate ontologies are the main challenges in building a skills management system. Manual ontology development can be characterized as an iterative, incremental and evaluative process. In the beginning, we provided the domain experts with a simple top level ontology to give a better understanding of the domain to be covered by the ontology. According to an initial baseline methodology (see Chapter 3), we advised the experts to use simple but helpful design rules, such as reducing

```
class-def Skills
    slot-constraint HasSkillsLevel cardinality 1
slot-def HasSkills
    domain Employee
    range Skills
slot-def WorksInProject
    domain Employee
    range Project
    inverse ProjectMembers
slot-def ManagementLevel
    domain Employee
    range one-of "member" "head-of-group" "head-of-dept" "CEO"
class-def Publishing
    subclass-of Skills
class-def DocumentProcessing
    subclass-of Skills
class-def DesktopPublishing
    subclass-of Publishing and DocumentProcessing
instance-of GeorgeMiller Employee
related HasSkills GeorgeMiller DesktopPublishing3
instance-of DesktopPublishing3 DesktopPublishing
related hasSkillsLevel DesktopPublishing3 3
```

Figure 12.5   A glimpse of the whole ontology

the degree of branching by setting a maximum of 5–10 branches, or limiting the maximum depth of the ontology. Then, domain experts independently filled their specific domain area within this top level ontology.

Using the design rules resulted in an overall reduction of the concepts which was a welcome side effect. In total, this step resulted in an ontology with more than 1000 concepts, including many duplicates. We then discussed and froze the ontology layer by layer, thereby identifying and eliminating some semantic duplicates in the ontology. Moreover, parts of the ontologies were restructured and apparently missing concepts were added.

For the development process we chose the brainstorming and mind mapping tool *MindManager* from Mindjet. We created concept hierarchies, reorganized them using simple drag and drop mechanisms, and applied the export function to make the ontology public on the web for review purposes. In addition, we annotated ontology elements with symbols or short notes about decisions that were made. For instance, a question mark denotes an open topic to be discussed, while a tick stands for an approved part of the ontology (see Figure 12.6). For group discussions we made large printouts of the ontologies and put them on the wall. The group of developers could view the current state of the ontologies with the meta data describing the state of the discussion. Then the unclarified points of the ontology were discussed and the ontology was rearranged and completed step by step. This approach to ontology development proved to be very successful concerning the outcome, the time required and the satisfaction of the ontology developers (see Chapter 3).

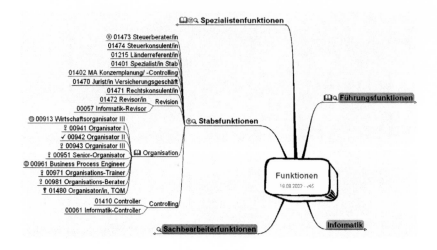

Figure 12.6 Meta-data for ontology development (screenshot from MindManager)

The iterative approach as sketched above makes it very difficult to get a clear versioning of the ontologies. Since most of the decisions are an outcome of a discussion, part of the changes never physically exist as a version of their own. It is also very hard to record the arguments that led to a decision without making a detailed protocol of the discussion. As this is more or less impossible due to the dynamic nature of the discussions, we only documented the result of a discussion and the main arguments for the decisions but left out any intermediate parts of the decision process.

While *MindManager* is an excellent tool to develop hierarchies in a cooperative brainstorming process, it does not offer real editing functionality. It does not check for duplicates in the ontology, relations can not be restricted in any way (e.g. range, cardinality), nor does it distinguish between the identifier for a concept and its representation. Such a distinction is a prerequisite for the construction and maintenance of multi-lingual ontologies which are a basic requirement in an international company, such as Swiss Life. All these features are supported by the ontology editor *OntoEdit* (see Chapter 7). Therefore, a combination of both tools might be close to a perfect ontology development tool: *MindManager* for the early development phase while using *OntoEdit* for extensions and maintenance.

12.2.5 Querying Facilities

The first version of the skills management system we developed at Swiss Life aims at finding people with certain skills and at supporting project staffing. Therefore, at the core of a query facility is the skills ontology from which a user can select terms to formulate a query. This kind of query simply interfaces to a corresponding RQL query to the Sesame repository (see Chapter 5). Likewise, a query that combines ontology terms by Boolean operators like AND and OR, can be translated directly into an RQL query.

More advanced queries combine ontology terms with arbitrary terms for a full text search over some or all of the information categories found on the home pages. Examples of such categories include projects worked on, special interests, former affiliations, etc. A typical home page would be divided into sections corresponding to some or all of the categories, with each section containing free text. To make sure that free text search terms can be restricted to one of the information categories, the division of home pages into sections is documented by the RDF annotation of the home pages. Thus, the free text search results can be filtered such that only those pages are returned where the search terms appear in the correct information categories.

For example, if someone is looking for a recommendation for Java training courses, the skills management system is used to search for 'Java' and maybe 'course'. An unspecific full text search will deliver all pages where the two search terms occur but it will not be able to recognize whether Java occurs in the 'Recommended Courses' section. By interpreting the RDF annotation, it is possible to filter out all pages where Java occurs in another section of the home page. For this the user has only to qualify the search term 'Java' as belonging to the information category 'Recommended Courses'. The search engine *QuizRDF* provides this functionality, that is, is capable of combining an ontology-based search (by interpreting RDF facts) with a free text search (see Chapter 8 for a detailed discussion of QuizRDF).

In order to achieve a match between an employee's skills description and some search terms entered, a down-posting along the concept hierarchy can be done, thus giving more flexibility to achieve a match. For example, if someone states that she has good skills in Java a full text search for 'Object-Oriented Programming Language' skills would not find this person because the term 'Java' does not occur in the home page. By down-posting, the search engine extends the query to include all subconcepts of 'Object-Oriented Programming Language' as well, thus also covering 'Java'.

Alternatively, in the case where a query does not give a satisfactory result, an up-posting might be appropriate. For example, when searching for an expert in Oracle 8 does not give a result, the query engine can automatically modify the query by substituting the query term with a more general term, for example, 'Oracle'. Another heuristic is to go up one step in the concept hierarchy and then include all immediate subordinate concepts. For example, if someone is looking for skills in 'XML' but no one has entered this skill, all employees with 'SGML' would be a good guess to ask, because 'XML' and 'SGML' have the same superconcept 'Markup Language' in the ontology.

A ranking of the results, as usually done by search engines, is problematic in our case. Of course, it is possible to define some similarity measure that takes all aspects into account, but this is not enough. Additionally, a user must be able to understand how the ranking was obtained. Moreover, it is unclear what is really meant when an employee specifies that he has a certain skill with a certain degree. If then another user looks, for example, for a specialist in Java, is somebody with basic knowledge in Java more relevant than somebody who is an expert in object-oriented programming languages? And what about somebody who has a strong experience with C++. Is he preferable to the other two? Only the user who posed the query can decide this.

Another problem concerning unclear semantics is that the meaning of the relationship between a concept and its sub-concept in the concept hierarchy does not correspond to the common understanding of *is-a*. For example, stating that somebody is an expert in Java does not imply the statement that the same person is an expert in object-oriented programming, and even less does it imply that the person is an expert in programming.

Our solution to the above-mentioned problems of producing a meaningful and intuitive display of search results is to move away from the traditional ranked list of relevant documents and to instead generate hierarchically displayed output (taking up an idea present in some search engines; see Chapter 9). This output visualizes the query results with numbers attached to relevant fragments of the concept hierarchy, each number indicating the skills level an employee has with respect to the skills concept it is attached to. We can then leave it to the user to decide which results are more relevant for the current search problem. This advanced query interface is currently under development.

12.2.6 Evaluation and Outlook

We are currently evaluating the existing version of SkiM in a pilot phase with more than 100 users. We found them to be very open to such a system and willing to publish their skills, provided they are publicly visible in the company. Most users said that they would not participate if their skills would only be seen by a few managers and a small group of people in the HR department. This confirms our hypothesis that employees will voluntarily

```
class-def Course                          class-def JobFunction
  subclass-of Education              slot-def RequiresSkills
slot-def ResultsInSkills                domain JobFunction
  domain Course                           range Skills
  range Skills                      slot-def HasJobFunction
slot-def HasEducation                   domain Employee
  domain Employee                       range JobFunction
  range Education
slot-constraint HasJobFunction cardinality 1
```

Figure 12.7   Sketch of an extension to the ontology shown in Figure 12.5

participate in such a system if it leads to higher visibility in the company which may lead to increased acknowledgement and better career opportunities.

Many users complained that browsing the skills ontology is too cumbersome and that it is difficult to find the right terms. A main emphasis of our future work will therefore concentrate on how to help users find appropriate ontology concepts. This might imply a redesign of the interactive user interface to improve navigation, or might call for an additional search functionality to find concepts by string pattern matching.

So far, the three ontologies for skills, education and job function have been developed independently from each other. In fact, however, there are dependencies which must be made explicit in the ontologies when more advanced functionalities will be realized in a future version of SkiM. For example, a certain education (including trainings and courses) results in certain skills, while a certain job function requires certain skills. A skills management system which takes care of such dependencies needs an ontology which describes them. For skills management tasks like career development and gap analysis, we consider it necessary that these dependencies are modelled in the ontologies. A sketch of how such an extended ontology could look like is depicted in Figure 12.7.

12.3 Automatically Extracting a 'Lightweight Ontology' from Text

12.3.1 Motivation

Swiss Life has moved from the local accounting standards used in individual countries to the International Accounting Standard (IAS) of the European Union. To support this move Swiss Life IAS specialists wrote the Corporate Accounting and Reporting Manual (CARM) which contains a description of the IAS insurance-specific accounting rules for European countries. Accountants all over Swiss Life use the manual to look up the information they need. However, it turned out that it is quite difficult to find relevant text passages in this 1000 page document. This is because CARM uses a very strict terminology which is generally unknown to Swiss Life's accountants in the various European countries because they are used to their country-specific accounting terminology and not yet totally familiar with the IAS. Of course, not entering the proper terms in a free text-based search engine leads to a poor search result.

To improve searching of the CARM document, Swiss Life started a case study with the aim of developing a search facility that supports a user in finding the right query terms. To this end, a special CARM ontology was built which contains the most relevant concepts from the document. The ontology is used by the search facility to map the terms of an initial free text query to the semantically most strongly associated ontology terms. This list of ontology terms can then be used by the user to modify the initial query. The benefits of this ontology-based search facility is a faster and easier access to the proper passages in the CARM and – as a side effect – an opportunity to get a deeper understanding of the new IAS vocabulary.

The main problem to be solved in realizing such a query support system is to build the ontology to be used. This is therefore the main issue dealt with in the following sections.

12.3.2 Automatic Ontology Extraction

The ontology development for the CARM search facility was severely restricted by the lack of time of the accounting experts. As a consequence, we had no expertise to develop the CARM ontology manually and were forced to follow a mostly automatic approach, using the experts only to correct and verify the outcome. Furthermore, time and financial considerations were additional reasons for an automatic approach because a manual development process takes much more time due to the negotiation and agreement process between the experts involved after each development cycle (see Chapter 3).

On the other hand, the CARM standard with its strict and coherent vocabulary offers good conditions for an automatic ontology development approach. Additionally, practically no synonyms and homonyms occur in the text that could confuse an automatic ontology extraction tool. The tool we used was OntoExtract, which takes one or more training documents as input and generates a draft ontology from them (see Chapter 6).

OntoExtract identifies the most relevant concepts in a document and infers weighted semantic associations between them. Consequently, the result cannot really be called an ontology yet, but would rather be a useful basis for a subsequent manual development into a fully fledged ontology. As we have not been able to add the manual step, due to the reasons mentioned above, we use the term 'lightweight ontology' for the resulting structure. In fact, it is an associative semantic network, which despite its simplicity, is the basis of a very useful application.

Clearly, a disadvantage of the automatic approach using a tool like OntoExtract is the small variety of relation types as compared to a manual development process. OntoExtract is currently not able to extract a concept hierarchy with class/sub-class relations, which is the basic relationship between concepts. However, as there are multi-word terms among the concepts extracted by OntoExtract, sub-class relations can be inferred on a syntactical basis via a sub-string match. For example, 'baby bond' and 'zero-coupon bond' can thereby be identified as specializations of the single-word term 'bond'. (Of course, such heuristics need to be used with care: 'software agent' and 'estate agent' would not be considered as sub-classes of the same class in most ontologies, for example).

The relations extracted by OntoExtract are weighted and take a value between 0 and 1, where 1 is the strongest relationship and 0 is no relationship at all. Table 12.1 shows an example of related concepts as generated by OntoExtract. The concept 'account' has 48 relations to other concepts, but only a few are depicted here. A relation describes a semantic closeness between two concepts, for example, 'account' and 'asset' which are strongly related in the CARM document. From the weighted relations, the CARM search facility generates the list of concepts most strongly related to a search term entered by the user.

The entire CARM was used by OntoExtract as a training document to develop the ontology. We also tried to use only parts of the document as training input but the outcome was poor. It turned out that although sufficiently

Table 12.1   Example of weighted concept relations

Concepts related to 'account'	With the relation weight of
Asset	0.86
Value	0.804
Amount	0.799
...	...
Maturity	0.054
Share	0.051
Security	0.051

large parts of the document mention most of the relevant concepts, the relations between concepts are incomplete as compared to acquiring the relations from the whole document. Even if we only take the relations with the strongest association weight into consideration, this still holds true.

Conversely, the problem that the extremely high connectivity of the concepts result in an ontology where every concept tends to be connected with every other concept was also encountered. The reason for this is the relatively even distribution of the accounting terms throughout the document so that there are no passages which use parts of the ontology predominantly. Applying such a densely connected ontology for suggesting additional query terms would result in a very low discriminative power among the concepts to be selected. This either leads to too many related concepts or to a few very strongly related ones that are trivial and of no use at all. The problem was solved by deleting those relations from the ontology which have a higher weight than 0.95 or a weight below 0.05. The strongest relations are removed because they occur with nearly every concept and thus are useless. The weakest relations are deleted because they are simply irrelevant. These cut-off values have been determined experimentally and proved to fulfil our application requirements best.

The extracted ontology was only superficially verified by the domain experts due to their lack of time, but concepts that were obviously irrelevant to the accounting domain could be eliminated. Thus, the ontology evaluation was done as part of the usability evaluation of the application.

The characteristics of the generated ontology are shown in Table 12.2. The number of actually extracted concepts depends on the minimal cut-off value set within OntoExtract. This can be implicitly adjusted by the ontology developer by saying how many concepts should be extracted. For the CARM search facility, 1500 concepts were generated, which was determined to be a reasonable number of concepts given the size of the CARM. An exhaustive extrac-

Table 12.2   Characteristics of the extracted lightweight ontology

Characteristic	Value
Number of concepts	1500
Number of relations	47086
Average number of relations per concept	31.39
Minimum number of relations for any concept	2
Maximum number of relations for any concept	64
Average relation strength over all concepts	0.2855

tion of concepts by OntoExtract, that is, without any cut-off value on concept weights, would give about 16,000 concepts.

12.3.3 Employing the Ontology for Querying

The extracted ontology was used in a search facility to support a user in finding the appropriate terms for a free text search on the CARM. The user starts by entering several search terms to specify what he or she is looking for (cf. Figure 12.8). If the result of the query (field 4 in Figure 12.9) satifies the user requirement, no further processing is needed. Otherwise, the user is motivated to invest some more time into reformulating the query. To facilitate this, the results page shows a list of concepts that are most strongly related to one of the query terms (field 2 in Figure 12.9). The user can then select from that list to reformulate the original query (field 1 in Figure 12.9). The user can also enter any new query term (field 3 in Figure 12.9). The results page for the reformulated query again shows a list of related concepts so that the reformulation step can be repeated as often as the user likes.

We originally considered two more possible scenarios for setting up the search facility. In the first of these, the user would start by browsing the ontology and picking up relevant query terms. We rejected this approach because a preliminary field study showed that users are not willing to invest much time in formulating a query. Entering a whole phrase or even a sentence was out of the question. However, after being confronted with an unsatisfactory result, users are much more motivated to invest further time.

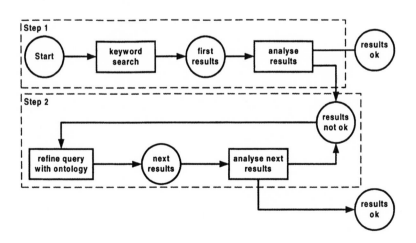

Figure 12.8   The query (re-)formulation process

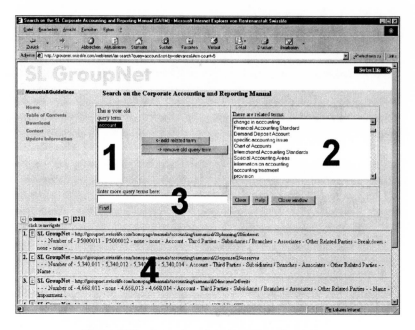

Figure 12.9   Query interface for query reformulation

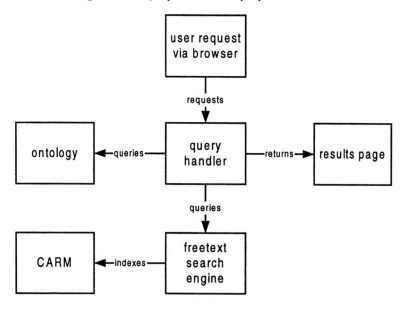

Figure 12.10   Architecture of the CARM search facility

Instead of offering a list of relevant concepts for query reformulation, we also considered showing for each original query term a graph of ontology concepts to which it is related. At first glance this might seem a better approach because it visualizes the structure and connectivity of the ontology concepts, but in our case users found such a richer interface confusing.

The architecture underlying the search facility is shown in Figure 12.10. The main component of the search facility is the query handler which gets all incoming user requests from the browser. The query handler calls the search engine for doing the full text search on the CARM and processes the returned results. In addition, the query handler matches the query terms to the ontology and merges the list of related ontology concepts with the results from the free text search engine to generate the output page.

12.3.4 Evaluation and Outlook

The evaluation of the application consists of two parts: the evaluation of the user interface and the evaluation of the underlying ontology. The usability of the search facility was evaluated with a questionnaire. The obtained user feedback led to some minor changes to the user interface.

The evaluation of the ontology consisted of two parts:

- compare the ontology with an existing glossary for the CARM;
- monitor the users' behaviour.

The results of comparing the ontology with the CARM glossary are given in Table 12.3. From the 290 glossary terms, 259 occur in the CARM document itself. This is somewhat odd but shows the deficits of a manual ontology (or glossary) building. There are 253 ontology concepts which contain at least one glossary term (more exactly, the names of those ontology concepts include a glossary term as a sub-string), while 85 glossary terms occur identically in the

Table 12.3 Comparing the ontology with an existing glossary

Characteristic	Value
CARM glossary terms	290
Glossary terms within the whole document	259 (89.3%)
Identical glossary terms and ontology concepts	85
Concept terms containing one or more glossary terms	253
Concepts within the whole document	1500
Concepts related to a glossary term	1178

Table 12.4 Users' query refinement behaviour (over a period of 7 months)

Characteristic	Value
Total number of queries	289
Total number of refined queries (1)	239
Total number of refinement steps (2)	301
Average refinement steps (for refined queries) (2)/(1)	1.26

ontology. The CARM ontology consists of 1500 concepts, which, of course, are all contained in the document. From these concepts 1178 have a relation to a glossary term. Thus, the ontology concepts are situated around the glossary terms, which can be considered as describing the terminological core of the underlying domain. It can be seen that the ontology and the glossary are well aligned.

The monitoring of user behaviour (cf. Tables 12.4 and 12.5) covers the usage of query terms, identifies if they are within the ontology or not, and keeps track of which terms are added and which terms are removed during query reformulation. The results of the monitoring gives an indication of the usage of concepts. From these results a decision can be derived as to which concepts and relations in the ontology should be deleted, updated (re-weighted), or inserted. Terms which were not used over a long period are candidates to delete, while terms that are often entered as query terms but are not in the ontology should be added to it. The number of times a query term is

Table 12.5 Query match with ontology (over a period of 7 months)

Characteristic	Value
Total number of queries	289
Single word queries	191
Single word queries in ontology	109
Single word queries not in ontology	82
Multiple word queries	98
Total match with ontology as a phrase	24
Total match with ontology as single words	32
Partial match with ontology	40
No match	2

replaced with the same concept is an indicator for increasing or decreasing the relation weight between both.

The evaluation results show that the ontology-based query refinement facility is well accepted by the users. Additionally, the following observations were made:

- Sometimes, users are looking for account numbers by typing in the number directly (12% of the queries). This suggests that a query answering system based on extracting facts from the tables in the CARM might be a highly welcome additional support. Instead of delivering text passages, such a system could give direct answers, for example, the definition of the entered accounting number, the accounting group to which the accounting number belongs, etc.
- Users often enter quite general terms as a starting point for their search. It might therefore be worthwhile to have a concept hierarchy in the ontology to better help the users in finding more specific terms.

Currently ongoing work aims at offering means to formulate more focused queries. An extension of the OntoWrapper tool (see Chapter 6) allows analysis of the structure of tables in the HTML code of the CARM document. Once that structure has been learned, it is possible to automatically index the contents of table cells, for example, as being numbers of accounting groups, or as being numbers of a sum account. The indexing is done by adding RDF annotations to the document. By substituting the currently used search engine with QuizRDF (see Chapter 8), a user will become able to qualify a number entered as having a certain meaning, for example, as identifying a sum account, thus posing a restriction on text passages that qualify as a query result. Further kinds of annotations might be worthwhile adding in order to achieve more focused search behaviour.

12.4 Conclusions

Both case studies evolved into real projects within Swiss Life. The CARM search facility is routinely being used by accountants from all business units of Swiss Life. The skills management system was rebuilt to meet the requirements of a productive system and is currently being used by over 150 employees on a completely voluntary basis. We are currently discussing with corporate and local HR departments enterprise-wide deployment of the system and how it would best fit into an overall knowledge management strategy.

During the case studies we encountered three main problem areas. One was the lack of available domain experts, who are needed to build an ontology. We had this problem with the case study on searching the CARM document, and solved it by doing an automatic ontology extraction from the document. In that case this worked very nicely because the document is based on a well-controlled vocabulary. In the general case, however, we anticipate that ontology extraction would be much more difficult. Moreover, we were only able to extract what we called a lightweight ontology (in fact an associative semantic network). In the general case, this network could form the starting point for a manual ontology development phase.

The second problem area encountered concerns ontology evaluation. It was found that within a certain range it can be determined which concepts, which attributes and which relations are OK and which should be re-examined, but beyond that any evaluation of the ontology becomes very difficult. The best way to do a further evaluation is via the application(s) for which the ontology was built. User response to the application dictates what is an appropriate ontology and what is not. Thus, ontology evaluation boils down to a large degree to application evaluation.

The third problem area became mostly visible within the skills management case study. We found that users have difficulty finding the right concepts within an ontology of several hundred concepts or more. This applies for selecting concepts for annotation as well as for querying. The problem of searching a large search space can be more easily solved by using ontologies. But then the problem of searching the ontology space needs to be solved. This is much simpler because the search space is smaller but it is nevertheless quite often a problem that usually gets only 'solved' by drastically cutting down the size of the ontology. This is clearly not satisfying. Interfaces such as those provided by QuizRDF (see Chapter 8) which allow users to locate themselves in an ontological space via a simple keyword search and then to browse the ontology from that point may help in this regard.

Notwithstanding the problems encountered, as we have discussed, the use of ontologies in the two application areas described clearly offered advantages over the use of traditional free text search technology and has led to systems which are now in live use.

13

Field Experimenting with Semantic Web Tools in a Virtual Organization

Victor Iosif, Peter Mika, Rikard Larsson and Hans Akkermans

13.1 Introduction

How do we test Semantic Web tools? How can we know that they perform better than current technologies for knowledge management? What does 'better' precisely mean? How can we operationalize and measure this? Some of these questions may be partially answered by simulations in lab experiments that for example look at the speed or scalability of algorithms. However, it is not clear in advance to what extent such laboratory results carry over to the real world. Quality is in the eye of the beholder, and so the quality of Semantic Web methods will very much depend on the perception of their usefulness as seen by tool users. This can only be tested by carefully designed field experiments. In this chapter, we discuss the design considerations and set-up of field experiments with Semantic Web tools, and illustrate these with case examples from a virtual organization in industrial research.

13.2 The EnerSearch Industrial Research Consortium as a Virtual Organization

EnerSearch is an industrial research consortium focused on IT and energy. Its aim is to create and disseminate knowledge on how the use of advanced IT will impact the energy utility sector, particularly in view of the fact that this industry branch is being liberalized across Europe. This liberalization is leading to strong market competition, supplier choice for customers, and major changes in industry structure such as mergers and acquisitions that cut across the traditional regional-national monopolistic borderlines. Thus, a hot issue for energy utilities is how they should strategically position themselves in a new deregulated market environment. Apart from the new market forces, new developments in both IT and energy technology are also an important factor in this strategic positioning. Innovations in energy technology increase the importance of decentralized local production and storage (e.g. 'green' renewable energy) leading to the concept of a 'distributed utility'. Advances in networked IT strengthen this trend, because they make it much easier to manage distributed utility operations at lower cost, but also open up the possibility of delivering novel energy e-services, and thus getting closer to the customer, for example, via the Internet and the Web.

In addition to the Web becoming smarter (which is the Semantic Web effort), it will also become more universal in the sense that it will not just connect computers, but essentially any device. This is variously referred to as 'ambient intelligence', 'universal connectivity' or 'ubiquitous computing'. Mobile commerce applications are one step in this direction, but basically all equipment, including home appliances such as personal audio and video, telecom and home control systems, and even heaters, coolers or ventilation systems, will become part of the Web. This development of smart networked equipment enables a broad spectrum of e-services for end consumers also in the utility sector, such as e-billing, energy management information and real-time feedback, energy saving advice-at-a-distance, home security and distant control, usage of the power line as a communication medium (e.g. for home networking), and smart buildings that are able to manage themselves.

EnerSearch carries out research projects to investigate such innovative possibilities. Some of this research is on innovative IT, especially on distributed intelligence. For example, EnerSearch has developed intelligent agents (so-called *HomeBots*) for smart buildings (Gustavsson, 1999), and electronic market-based control for applications such as distributed load management and in-building comfort management (Ygge and Akkermans, 1999; Carlsson et al., 2001). Such applications are also field tested (see Figure 13.1). All of these new imagined e-services are technically challenging, but will also require and induce different behaviours and attitudes from the adopting end consumers as well as from the businesses delivering these e-services. Therefore, EnerSearch is also carrying out business modelling and marketing research (Larsson et al., 1998, 1999; Sweet et al., 2000; Sweet, 2001).

Figure 13.1   Field test sites of EnerSearch work. Top, Villa Wega, Ronneby, Sweden; bottom, smart office building at ECN, Petten, The Netherlands. ECN is one of the shareholding companies in EnerSearch

EnerSearch has a structure that is very different from a traditional research company. Research projects are carried out by a varied and changing group of researchers spread over different countries (Sweden, United States, Netherlands, Germany, France). Many of them, although funded for their work, are not even employees of EnerSearch. Thus, for its knowledge creation function EnerSearch is organized as a virtual research organization. The insights derived from the conducted research are intended for interested utility industries and their IT suppliers. Here, EnerSearch has the structure of a limited company, which is owned by a number of companies in the industry sector that have an express interest in the research carried out. Shareholding companies include large utilities in different European countries including Sweden (Sydkraft), Portugal (EDP), The Netherlands (ENECO, ECN), Spain (Iberdrola), and Germany (Eon), as well as some worldwide IT suppliers to this sector (IBM, ABB). Due to this wide geographical spread, EnerSearch has the character of a virtual organization also from the knowledge distribution point of view. In addition, for the general public interest it maintains a website (www.enersearch.se) where it publishes many of its research results as papers, reports and books. Thus, dissemination of produced knowledge on IT and energy is a key function for EnerSearch. Within the On-To-Knowledge project, EnerSearch investigates whether Semantic Web methods and tools might be helpful to improve on this function, especially focused on its web information provisioning.

13.3 Why Might Semantic Web Methods Help?

The EnerSearch website happens to be an important mechanism for knowledge dissemination. In fact, one of the shareholding companies actually entered EnerSearch directly as a result of getting to know it from the web. Nevertheless, the information structuring of the website leaves many things to desire. It has a very traditional structure typical for many research organizations: some central 'about us' information, an overview of what projects have been done, and what researchers are involved. From there one finds the many electronic papers, reports and books that contain the EnerSearch research results in various knowledge areas. So, the real entry to what constitutes the heart of the EnerSearch – insights on how advanced IT may impact the energy business in a deregulated market environment – is along the lines of projects and authors. Clearly, this is very much a website organization driven by the information suppliers – projects and authors. Consequently, it does not adequately cater for the demand-side needs of information seekers – for example, users from interested utility companies. They are generally not at all interested in knowing what the projects or who the authors are, but rather in finding answers to questions that are important in this industry domain. For example: Does load management lead to cost savings; if so, how big are they, and what are the upfront investments needed? Can powerline communication be technically competitive to cable modems or ADSL? Is there an interest in the residential market in new energy services? Et cetera.

The main difficulty with the current EnerSearch web server is that it is rather hard to find the right information on such issues (unless you know something about the projects/authors, or take some more time to browse and search). This is the result of a supply-side (or inside-out) rather than demand-side (or outside-in) style of information provisioning. Also, the EnerSearch search engine supports free text search but not content-based search. This standard text search has the usual problems: lack of precision and retrieval of irrelevant information resources. One possibility is to completely restructure the information on the website, but this is a huge burden on the information providers – researchers for whom the project/author structure is quite natural and, moreover, in a virtual organization it is difficult to impose that they spend time on what they will mostly see as additional overhead time. Therefore, an alternative solution is not to change the structure of the website too much, but to make it more accessible to users, by providing an extra information access viewpoint that is structured such that expected demand-side questions and interests are the starting point instead of the end point.

This is where Semantic Web methods and tools might clearly help. There are several reasons for this. First, they might help, as already pointed out, to limit the time that is spent by the information suppliers. Furthermore, the target groups of the web information are quite clear: researchers in the field, staff and management of utility industries, and to a lesser extent, a general outside audience that happens to have an interest in IT and energy, smart homes, etc. Accordingly, it is quite possible to form a clear picture of what kind of topics and questions would be relevant to these target groups. Finally, the knowledge domain in which EnerSearch works is relatively well defined. So, it should be certainly possible to come up with a domain ontology, describing the various important topics and their relationships, that is of good quality and fairly stable. In actual fact, some of the semantic sitemaps shown later are based on a lightweight 'EnerSearch lunchtime ontology' that took the two developers no more than a few hours (over lunchtime) to develop.

13.4 Design Considerations of Semantic Web Field Experiments

Thus, Semantic Web tools and ontology methods might help EnerSearch with its internal and external knowledge management. For this reason, EnerSearch is conducting a case study to implement and evaluate methods of semantic access to its website (Iosif and Ygge, 2001, 2002). There are several independent elements that are to be considered in general when it comes to the design of field tests of Semantic Web tools:

- different *modes* of information gathering that are employed by users when trying to find the information they are looking for;
- different *target user groups* that are to be distinguished from the viewpoint of differences in interest, motivation, the nature of information they pursue, and background knowledge they bring;
- different *personal styles of information use* that relate to differences in individual cognitive styles in handling information.

13.4.1 Different Information Modes

If we want to evaluate semantic methods, we first have to establish a clear reference point for comparison, or some kind of unit in terms of which we can measure the quality of new ways of doing things. In the present case, the reference point is the usual web way of gathering information: keyword search and hyperlink-based browsing. This provides the yard stick with which we can measure how good novel semantic approaches are. Here, we consider two different semantic web-style information modes: (i) ontology-based search; (ii) ontology-based browsing. We provide some examples of both these semantic information modes later.

13.4.2 Different Target User Groups

In general it is highly important to carry out a proper segmentation of target users, because often they are quite different, each bringing with them their own view or, so to speak, 'ontology' regarding the information they seek. These differences will clearly influence their perception of how good semantic web methods work. In the case of EnerSearch, some target user groups and their characteristics are fairly easy to identify, namely (i) externally: interested staff and management in utility companies; (ii) internally: researchers involved in IT and energy projects. These two groups are very different. Utility staff is generally more interested in business issues. Even when it concerns new technology, the interest will not first of all be with the technicalities, but more with envisioning the potential applications ('what can we do with it?') and establishing the business case for it ('what are the benefits?'). For researchers, it is more or less the other way around. Refereed publications containing a strong and technically detailed, usually formal, line of argument are key to academics. In contrast, in the utility business, technical competence has always been seen as very important, but research has always been limited: a typical figure for R&D spending in the utility industry is no more than 2% of total annual revenue. This figure tends to go down with market deregulation: cost cutting is crucial, and the strategic time horizon decreases. In addition, IT is typically not a core competence in utilities that tend to have a very much 'cable and wire' view on networks, a view that is carried over also to IT (so we may invite the reader to imagine how to explain convincingly the concept of 'software agent' in such a setting). Thus, the outlook of researchers and interested utilities on the same broad topic of IT and energy is significantly different. Last but not least, there seems to be (as measured by hits and downloads on the website) a third user group of the EnerSearch website that is much more diffuse, viz. the more general public that is unknown but interested in various bits and pieces of the research results and trends. Of course, adequately predicting the expectations and interests of this broad group for the purpose of adequate website information provisioning is very difficult, and herein lies an interesting challenge for application of Semantic Web methods.

13.4.3 Different Individual Cognitive Styles

Yet another dimension that will influence the outcome of Semantic Web tests and experiments lies in the fact that people differ in the personal style with which they handle information and make corresponding decisions. As this is a novel element in Semantic Web research, we will go a bit deeper into how cognitive styles of information use can be framed and measured. A useful theory to do this stems from the field of organizational behaviour and is called decision style theory (Driver et al., 1993). It has roots in Herbert Simon's theory of bounded rationality that says that humans often cannot process all relevant information pertaining to a decision to be made, but are content to take in a limited amount of information that is adequate to make a decision that is 'good enough'. This is called *satisficing* decision-making behaviour.

Decision style theory generalizes Simon's idea as follows. People differ markedly in how much information they use in decisions. There are people who seem to quickly jump to conclusions, and some who will not move without a thorough analysis. A person that is happy to consider the minimum amount of information needed to come to a reasonable, 'good enough' decision is called satisficing in his/her information use. In contrast, a person that wants to do a thorough analysis first to make sure that no relevant piece of information has been missed before coming to a decision, is called maximizing in use of information. This difference is illustrated in Figure 13.2. The first information you receive about a problem generally is most valuable to you as a decision-maker. When you know almost nothing about a problem, those first few pieces of information can be real eye-openers. Later, after you already have gained a lot of knowledge about a situation, new items of information are much less likely to add to, or alter, your understanding of the situation. The satisficing versus maximizing styles of information use are related to the perceived *value of information*, and constitutes one dimension of cognitive style. In addition to the question how much information you use to come to a decision, there is a second dimension called *solution focus*. Here, the distinguishing issue is that in decision-making, some people tend to employ information to come to a single definite course of action, while others tend to view

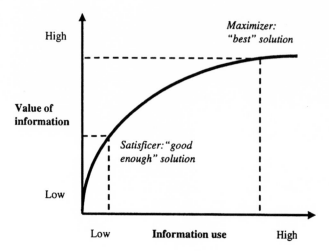

Figure 13.2 Two styles of information use, related to perceived value of information

information as leading to a variety of different possible solutions at the same time. These contrasting styles in decision-making focus are depicted in Figure 13.3 and are called unifocus versus multifocus.

Then, if we combine these two style dimensions – amount of information use and solution focus – we get a set of different individual information-based decision-making styles as presented in Figure 13.4. The ensuing decision styles are called decisive, flexible, hierarchic, integrative and systemic. The

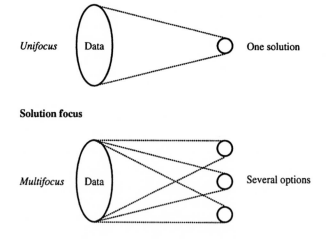

Figure 13.3 Solution focus dimension in individual decision styles

Satisficing **Information Use** Maximizing

	Decisive	Hierarchic	
Unifocus	Fast Action-oriented Efficient	Analytic Logical Quality	Systemic
Solution **Focus**	Flexible	Integrative	Analytic Comprehensive Prioritizing
Multifocus	Fast Action-oriented Adaptable	Analytic Exploratory Creative	

Figure 13.4   Matrix of decision styles (Driver et al., 1993)

decisive style is a satisficing and unifocus style. Decisives use a minimum amount of information to rapidly come up with a clear solution for a problem. Get-things-done-now is characteristic of this style. Decisive communication is short and to the point. Decisive personalities say what they mean and mean what they say. Flexible style people use moderate levels of information (as do decisives), but they interpret the same information in different ways at different times depending on the point of view they currently hold. They are able to quickly adapt to the expectations of others and so frequently change direction depending on who is present. Flexibles prefer not to plan, because planning makes them feel trapped and unable to exercise options. In contrast, the hierarchic style makes maximum use of all information to reach the best single conclusion. For hierarchics, there is clearly a 'best way' of doing things. High data use is dictated by the desire to ignore no detail that could diminish the quality of a conclusion. Once information is rigorously analysed – often in quantitative fashion – the resulting plan is complex but highly focused. The integrative style tends to process a large amount of information (like hierarchics), but they use it (like flexibles) to generate many solution alternatives. Integratives habitually produce a wide variety of possible interpretations of information about a situation simultaneously. Their plans are constantly open to change: to be seen as a good initial course of action, but not something to stick to indefinitely and thus to be revised when the evolving situation so requires. Finally, the systemic style turns up empirically in the quantitative studies done regarding decision styles (empirical data actually involve many tens of thousands subjects in many different countries). It combines qualities of both the integrative and hierarchic styles. It is a maximizing style in information use, but it is both multifocused and unifocused in decision making. Systemic planning is comprehensive. It is unifocused in terms of having a single overarching and unifying goal, but it is multifocused in that it will contain many different sub-goals and many tactics and strategies for hitting those targets.

Individual differences in information processing style thus capture essential characteristics of people in an organization, as to how they want information presented and how they typically communicate. Generally, people do not have a single style, but usually some mixture, with one or two being dominant. Furthermore, decision styles are not static. They change as a result of changes in environmental parameters such as work pressure creating cognitive (over)-load, or whether we are in a daily work situation or in a formal, self-aware setting (operating vs. role styles). On a longer time scale, the decision style of a person also may change. Styles are learned habits of doing things, and are influenced by the social environment we live in, the work we do, the demands and expectations that go with it, etc. As these factors change, our information-processing styles may change as well over the years. Many aspects of decision style theory cannot really be touched upon here, and for a comprehensive overview the reader may consult the literature (Driver et al., 1993; Larsson et al., 1998, 1999). To measure a person's decision style, standardized questionnaires exist that have been validated over many years of research. These measurement instruments are also used in the Semantic Web case study carried out by EnerSearch.

13.4.4 Hypotheses to be Tested

In sum, in designing Semantic Web experiments, different design dimensions are of importance: variations in information modes, in target user groups, and in individual information-processing styles. Finally, an experiment must be based on one or more clearly formulated hypotheses that can be verified or falsified, for example, by empirical-statistical methods. A possible list of testable hypotheses regarding Semantic Web-based information seeking is:

- *H1:* Users will be able to complete information-finding tasks in less time using the ontology-based semantic access tools than with the current mainstream keyword-based free text search.
- *H2:* Users will make fewer mistakes during a search task using the ontology-based semantic access tools than with the current mainstream keyword-based free text search.

- *H3:* The reduction in completion and number of mistakes will be more noticeable for less experienced users.
- *H4:* The reduction in time will also be more noticeable for users lacking a detailed knowledge of the underlying technical system implementation.
- *H5:* The ontology-based semantic access tools will be perceived as more useful than free text search by different types of persons for a broad range of domains and knowledge-acquisition scenarios.
- *H6:* The effort for developing and maintaining the ontology and information structure will not significantly exceed the effort to develop and maintaining the free text approach.

In a field experiment, hypotheses such as these are tested for their significance, and how their validity varies with different information modes, target user groups, and individual information-processing styles is investigated.

13.5 Experimental Set-up in a Virtual Organization

The case study that we are carrying out on Semantic Web tools under field conditions is to be seen as an exploration study. It does not constitute a comprehensive comparative study. For this we would need more information retrieval tools (both standard ones and ontology-based semantic access tools) and in addition a larger group of subjects to compare the results and extract statistically significant empirical conclusions. However, this exploratory case study will give a good indication as to what directions we should aim for semantic tools and methods in the future.

13.5.1 Selecting Target Test Users

The involvement of knowledge users in the experiment from the beginning is important because of the interaction between the users and tasks on the one hand, and their pre-knowledge, or lack thereof, of domain and/or systems on the other hand. We identified several types of target users for the tests through conducting a set of pre-trial interviews. As a result, the evaluation experiment includes three different types of interest groups, as outlined previously. One group consists of staff members from four different shareholder companies (the companies involved in the case study are: Sydkraft AB, Sweden, Iberdrola from Spain, Electricidade de Portugal, and ECN, The Netherlands). A second group consists of researchers from different scientific fields, several having at some time participated in EnerSearch projects. The third and final group intends to represent more or less a general outside audience and consists of students (studying at the Department of Software Engineering and Compu-

ter Science at the Blekinge Institute of Technology in Sweden). Finding
information on a personal basis is important for all of these three groups,
but for various reasons (that well reflect those found in geographically spread
virtual organizations) they are generally limited in their time to invest for
searching knowledge. The majority of the test users are familiar with the
EnerSearch web and have used it before. There were of course also those
who had never heard of the EnerSearch web but were introduced to and
instructed on how the EnerSearch web is functioning.

13.5.2 Tools for Test

The test people will use three different tools: the two ontology-based semantic
tools QuizRDF and Spectacle (described in Chapters 8 and 9, respectively)
and the free text search tool, the EnerSEARCHer. These tools and what kind
of results they produce are discussed in detail later. The ontology-based search
tool QuizRDF is chosen for the case study for two reasons. The tool has the
advantage that the user can start with simple queries consisting of only a small
number of search terms in order to get a picture of what kind of information is
available in the EnerSearch knowledge base. There is then a continuum
offered from the common keyword-based search to different levels of semantic search. The reason for using the ontology-based browsing tool Spectacle is
that it presents the information according to the inherent structuring that is
offered by the domain ontology. This gives a valuable semantic context for the
user not available through standard information retrieval tools. The information is presented in such a way that the path that leads to the information adds
to the users' understanding of the semantic role of the information. Each
concept chosen is 'surrounded' by other ontologically similar concepts.
Semantic generalizations and specializations (in different dimensions) are
also offered to the user when browsing for information.

13.5.3 Test Tasks and their Organization

An important issue is the choice of the application tasks posed to the test users.
For our purpose, they need to be simple enough to be learned in a short time
but complex enough to challenge the users in the test groups. An example of a
knowledge-acquisition task given to the test users is to find certain answers to
predefined questions about finding information in the EnerSearch knowledge
base. Significant effort has gone into setting up a useful set of predefined
questions such that they allow for adequate comparative analysis, and at the
same time, adequately sample the different interests of the test user categories
and cover the various available information tools in a representative fashion.

An experimental problem within the test case study is that if one gives a user the same exact search task to do with two different tools there will be most probably be a transfer error. This means that it is rather certain that they will be unlikely to repeat errors the second time they perform the task, and that they will remember how they did something and will not need to figure it out the second time around. To avoid this transfer effect, we designed three different but comparable scenarios, each involving the same kind of knowledge-acquisition task in the same domain but involving a different aspect of the knowledge base.

The test questions are divided into three blocks. There are six different ways to organize the tools in different order (Spectacle, QuizRDF, EnerSEARCHer) and equally many ways to organize the blocks of questions. In total there are 36 different ways to organize the order of the tools and the search questions. Each person will have a unique combination of tools and search questions. Through this permutation of question and tools order, we will make a fair and statistically sound evaluation of the tools.

To facilitate the test user's acquisition of the knowledge required to perform the scenarios and to ensure a uniform understanding of the domain across users, the application domain was explained to the test users by an email that was sent out in advance. The test users and the domain were especially chosen to avoid differences in the subjects' previous exposure to the domain. Of course, these points are not relevant when testing test users that are experts in a specific domain. A repository of knowledge bases and scenarios to test the tools that could be shared by different researchers makes it a better comparative evaluation among the different approaches, as well as reduces the amount of work required to evaluate the test. For every major ontology area a different scenario is used. It is important that the different scenarios are as comparable in size and complexity as possible for the results of the experiment to be meaningful.

13.5.4 Experimental Procedure

After having determined the kind of experiment to be carried out, our hypotheses, the type of users, and the nature of the search task, we are in the position to plan other details of the experiment. These include, for example, what information will be given to the test users and in what format, what kind of interaction can the test users have with the experimenters during the tests, for example, can they ask questions that are related to the domain or the tools used in the experiment. How many iterations or problems will be given to each test user and in what order, and an indication of the success criteria for the test user so they know when they have finished the scenario they are given.

The experimental set-up has to be carefully designed to control as much as possible the variables that could affect the outcome of the experiment. In this case, the order in which the tasks are executed might also affect the results, therefore we switch the execution order for different test users. We asked the test users to fill out a background questionnaire prior to the execution of the experiment to form the groups as balanced as possible. For all test users, their information-processing styles are also determined separately by administering the standard questionnaires that go with the decision style theory.

Our controlled experiments compare the EnerSEARCHer with the performance of test users using the QuizRDF and Spectacle tool. Each test user is confronted with different questions related to the EnerSearch web. Each scenario will be performed with an ontology-based tool and the free text search tool. In this way the results will be independent of the complexity of the scenario. Summarizing, by a careful experimental design, all factors are balanced to even out the qualifications of the test users for each one of the groups, the number of times that each tool is used for each scenario, and the number of times that each tool is used in the first place.

13.5.5 Determining What Data to Collect

The data collected during the experiment may be determined and/or limited by what is possible in terms of instrumenting the ontology-based search tools and the knowledge environment. Intrusive ways of recording data should be avoided. For example, we should not ask the users to fill out a long form to describe what they just did if that is going to disrupt their thoughts and if it makes the overall time to complete the task longer. The following data are collected during the execution of our scenarios:

- Time to complete the whole search task.
- Automatic log of changes performed to the knowledge system.
- Automatic log of errors in the search task.
- Automatic log of commands executed (e.g. add a new problem-solving sub-step similar to an existing one).
- Detailed notes of the actions performed by the test users (taken manually by the experimenters by means of onsite observations) including how users approach the given problem and what materials they consult. For this purpose we ask the test users to verbalize what they are thinking and doing during the execution of the scenario. We use tape recording during the interviews since we found that only note taking is not sufficient.
- Post-trial questionnaires the test users fill out at the end of the experiments, with questions regarding the perceived usability of the tools.

The data collected should be sufficient to test our hypotheses, as listed above, and confirm whether they are relevant or not. Collecting data is very useful because it not only proves/disproves the hypotheses, but it also helps to explain the outcome of the experiment, and to explore the potential causes of certain experiment outcomes. We found that conducting pre-test interviews was useful not only to help refine the actual evaluation set-up but also to set up the data collection strategy.

13.5.6 Evaluation Matrix and Measurements

Each person performing the experiment starts the test by a login procedure. The user's name and which company or organization the user belongs to is registered. By using login it will also be possible to see when and from where the user entered the EnerSearch web. We will be able to measure the time period the different users will spend on answering each question in the experiment. After each question the users are asked to explain how they received the information for answering the question. By collecting the information and comparing the user's answers to a question it will be possible to evaluate the different ontology-based semantic tools. It is also important to evaluate the level of appreciation for the different tools. The questions to evaluate this would be for example: Were you able to complete search tasks in less time using the ontology-based tools than with a standard free text search tool? Was it easy to use the ontology-based tool? Did you understand the interface? Were the answers correct? Was the answer relevant to your search question?

To be able to answer the questions in a reasonable time, the case study needs to be limited to a certain number of questions for each ontology-based tool. The case study is therefore restricted to a limited number of questions, in order to avoid user defection due to lack of time to complete the experiment. To avoid defection we have also made it possible for the users to log in at any time to complete the questionnaire.

It clearly follows that Semantic Web tool tests require a very careful experiment design. Prospective test user groups and test tasks must be carefully balanced to allow for adequate empirical-statistical testing of hypotheses that must be explicitly formulated in advance. Empirical data gathering in such an experiment must be rich, including various qualitative methods (such as pre- and post-trial semi-open interviews, collecting verbal protocols during the experiment, onsite observation), as well as quantitative methods (e.g. Likert-scale questionnaires, electronic logging of actions and execution times, and statistical processing of resulting data).

13.6 Technical and System Aspects of Semantic Web Experiments

13.6.1 System Design

The baseline hypothesis is that the ontology-based solution will make it easier for the users to locate the information they seek and will also make it easier to share knowledge with others in the organization. Users can choose between two new ways of finding information:

- querying the knowledge base with the ontology-based search tool QuizRDF;
- navigating and browsing the presentation built from the website knowledge base using the Spectacle presentation tool.

Field tests will evaluate the navigation and search facilities of this system against the traditional keyword-based EnerSEARCHer tool that has been previously in use for search within the domain. The complete design of the system for ontology-based information retrieval is shown in Figure 13.5. The system is built using tools and technologies developed within the On-To-Knowledge research project and thus it follows many of the design decisions taken within the project, such as:

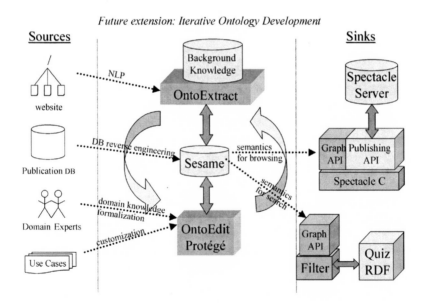

Figure 13.5 Overview of the system design for the EnerSearch case study

- a loose coupling of tools through HTTP;
- the use of a central storage facility (Sesame);
- support for weakly structured media (plain text, HTML and PDF formats);
- the choice of RDF(S) as a knowledge representation format;
- the conformance to a commonly agreed upon data model.

For the end user, the functionality of the system is made available through the HTML interfaces of the QuizRDF search engine and the Spectacle presentation that we present further below.

13.6.2 Ontology Engineering, Population, Annotation

The process of setting up the system begins with acquiring a base ontology by running the OntoExtract extraction tool (see Chapter 6) on the publications stored on the website of EnerSearch. The resulting lightweight ontology contains a shallow hierarchy of concepts and relations between the concepts, plus the classification of pages according to concepts. Once submitted to the central Sesame server, it is supplemented with the ontology compiled from the EnerSearch publication database by reverse engineering it into RDF(S). The OntoExtract domain ontology obtained through natural language processing (NLP) and the ontology describing the publications are naturally linked together via the URLs of the publications.

The second step is to refine the ontology by using the OntoEdit tool. As guidance in this process, the editor of the ontology consults domain experts and takes proposed use cases into account. When it is implemented, the ontology engineer will be able to feed back the resulting ontology as background knowledge for OntoExtract, realizing iterative ontology development in effect. (The background knowledge is also stored in a Sesame repository.) Such a feature will make it possible to transform the rather lightweight ontology obtained by NLP into a richer, more structured form while maintaining the advantages of an automated approach (e.g. scalable, automated classification).

In a third and last step, two presentations are generated from the ontological data for searching and browsing. Generating an ontology-based search engine is a one-click process: it simply requires selecting the Sesame repository to use. (Naturally, this assumes that the ontology conforms to the common data model.) Subsequently, the tool processes the pages that are annotated by the ontology and augments its keyword index with ontological data. The result is the tailored search engine described below.

As the iterative ontology development process was not available at the time of this writing, the ontology used in the case study is based upon a combination of the OntoExtract generated lightweight ontology and the converted publication database. The QuizRDF search engine, however, has been generated from a version of the lightweight ontology where only key concepts are present. This was strictly necessary because QuizRDF is a generic tool that has no understanding of the ontology beyond RDF Schema and parts of the data model used within the On-To-Knowledge project. Also, a different version of the publication database has been used in the case of QuizRDF for technical reasons, notably that QuizRDF can only search property values on the instance level.

Creating the Spectacle browsing interface is a programming task that requires intensive use of the Sesame client API, the Spectacle publishing API and some programmatic ontology transformations, for example, to infer new facts that are not explicitly stored in the ontology. For example, if a page is about a certain concept then it is also about the super-class(es) of the given concept. Once the content is generated, it is stored on the Spectacle presentation server where users can access it using a JavaScript enabled browser.

13.7 Ontology-based Information Retrieval: What Does it Look Like?

Ontology-based information retrieval will support the user when s/he is querying the knowledge base. It will allow the user to formulate queries in a more natural way, for example, by helping to select and combine relevant terms. Such an approach will also enable advanced types of queries that are not possible with non-ontology-based search tools, for example, by locating information when specifying the scope of interest or by providing help with generalizing or refining a query. Also, with the browsing approach embodied in the Spectacle presentation, it is expected that the process of information retrieval will also add to the user's understanding of the domain.

13.7.1 Ontology and Semantic Sitemaps

As sufficient technologies for enriching the ontology obtained by NLP are not available at the moment, the ontology employed in the case study is a lightweight ontology that was manually developed in a short time. We note that the domain is such that it lacks a well-defined sub-class hierarchy (rather, it is a network whereby only in some parts a hierarchy exists). However, we were able to capitalize on the strengths of an automated approach, namely efficient classification of the fairly extensive publication database complemented with the automated selection of key concepts and related concepts. Thus, the work imposed on the information providers was very limited.

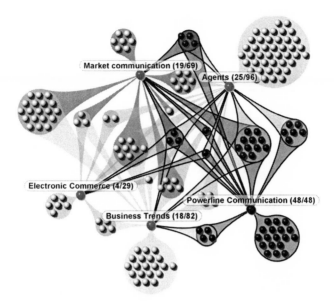

Figure 13.6   Fragment of a lightweight ontology showing some of the topics and sub-topics on which the EnerSearch website provides information

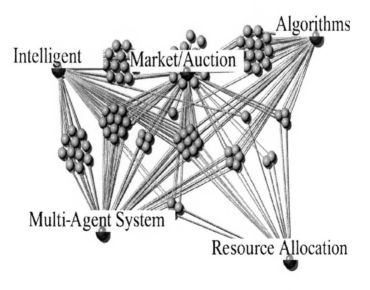

Figure 13.7   Semantic map of the EnerSearch website for the sub-topics related to the concept of 'agent'

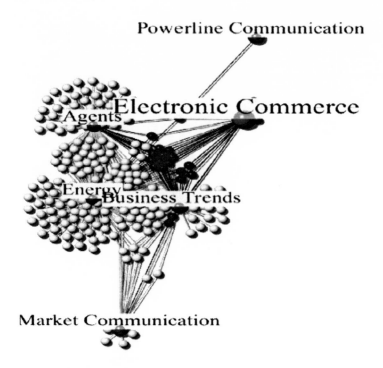

Figure 13.8 Semantic map of the EnerSearch website showing the overlap links between the topic 'Powerline Communication' and other main topics

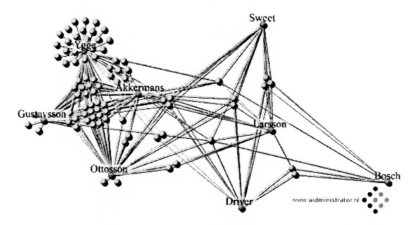

Figure 13.9 Semantic map of the EnerSearch website showing the semantic distance between different authors, in terms of their disciplinary field of research and publication

Previous experience has shown that lightweight ontologies can still provide meaningful insight when combined with automated approaches. To demonstrate that, Aidministrator's Webmaster tool (v1.4) was used with a baseline lightweight ontology, of which a fragment is shown in Figure 13.6. Outcomes from the automatic semantic sitemap analysis are shown in Figures 13.7–13.9.

Figure 13.9 provides interesting information about the relation between the different authors that only becomes clear due to the ontology-based semantics. The authors, Ygge and Gustavsson, have mainly provided technical papers whereas Sweet and Bosch have written mainly business administration oriented papers. Some authors have a more general interest (cf. Akkermans), and this is visible in the semantic map of the EnerSearch website that shows the semantic distance between the authors in terms of their disciplinary fields of research and publication.

13.7.2 Semantics-based Information Retrieval

For the information retrieval approaches (EnerSearcher, QuizRDF and Spectacle) that correspond to different information modes, examples of screenshots are shown in Figures 13.10–13.13.

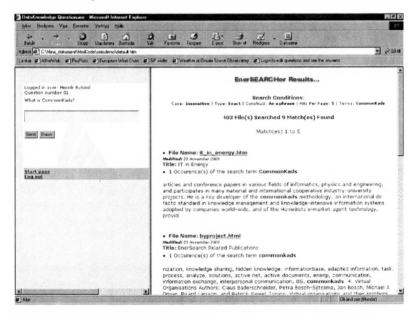

Figure 13.10   The integration of the search tool and the electronic questionnaire. In the left frame, a question is shown 'What is CommonKads?' In the right frame, a free text search result is shown

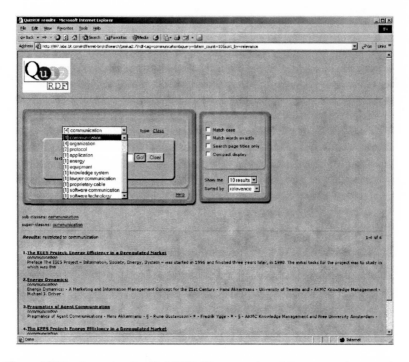

Figure 13.11 A search using QuizRDF. The user selected the concept 'communication' resulting in four hits. (Although two appear to be the same, the links point to separate parts within the HTML document.) The drop down list at this point shows other key concepts that appear on these pages. Selecting one of these results in further filtering of the selection

The EnerSEARCHer supports standard features like searching on words or combinations of words and using case sensitive and intensive search. The search quality of the EnerSEARCHer is similar to the quality obtained on most standard web-search engines. The difficulty with this search tool is that it returns a lot of irrelevant hits, can be rather sensitive to spelling and performs very poorly with synonyms. A screenshot from the EnerSEARCHer is shown in Figure 13.10.

QuizRDF first presents the user with a text entry box and a drill-down menu. The drill-down menu contains a list of all the resources types stored in QuizRDF's index, which are concepts and a 'Publication' class in our case. The drill-down menu gives the user the option to narrow down the search to a specific concept: after selecting a concept, only the pages related to that concept are shown. By selecting the class 'Publication' in the drill-down

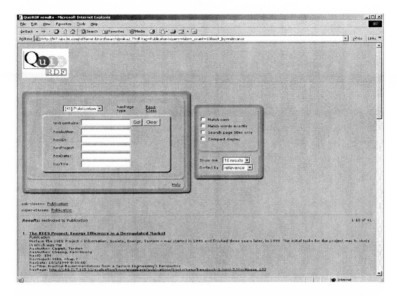

Figure 13.12   Here, the user selects the concept 'Publication'. Due to the rich set of facets attached to this concept, the user has the additional optional of searching on a publication's ID, author, title, date of publication and the project to which it belongs

menu, the user has the opportunity to search on the various fields of the publication database. At any time the user can enter natural language text into the text entry box as with free-text searching. QuizRDF combines these search criteria using the AND operator and matches the resulting query against its ontology-based index. The response is a list of QuizRDF resources ranked according to a resource's relevance to the user query. Screenshots from QuizRDF are shown in Figures 13.11 and 13.12.

Finally, using the Spectacle presentation, users can start browsing by selecting an item from the list of all concepts, key concepts, authors, projects or years of publication. Subsequently, the user can further narrow the search by choosing a co-occurring concept (i.e. one that appears along with the concepts chosen so far on some of the pages) or a related concept. This step is repeated until the list of pages (hits) is reduced to a certain number. A screenshot of Spectacle is shown in Figure 13.13.

13.8 Some Lessons Learned

To demonstrate the real value of Semantic Web methods we need to carry out field experiments. We have outlined a number of hypotheses that should be

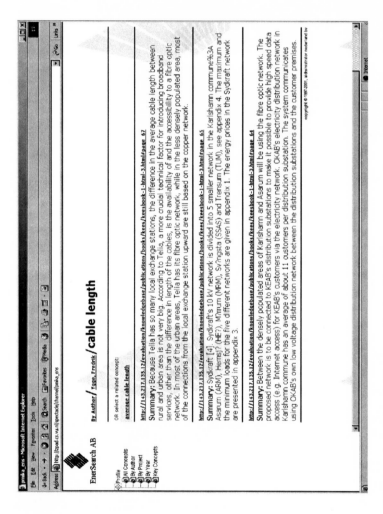

Figure 13.13 Semantic browsing with Spectacle. The user selected the 'By Author' option, then chose the author Fredrik Ygge and the concept cable length. The result is all the pages (co)authored by Fredrik Ygge where he discusses cable types

studied in representative case studies. We have also described what kind of variables have to be taken into account, how data collection, evaluation, experiment procedure, and system design can be done, and we have sketched the importance of the human side of information processing.

At the time of writing, we cannot yet give the final results of our field tests of Semantic Web tool use. However, several lessons and conclusions can be already be derived from our practical experiences so far.

- Semantic Web tool tests and case studies in the field require a very careful experiment design. Prospective test user groups and test tasks must be carefully balanced to allow for adequate empirical-statistical testing of hypotheses that must be explicitly formulated in advance. Empirical data gathering in such experiments must be rich, including various qualitative methods (such as pre- and post-trial semi-open interviews, collecting verbal protocols during the experiment, onsite observation), as well as quantitative methods (e.g. Likert-scale questionnaires, electronic logging of actions and execution times, and statistical processing of resulting data).
- The case study validated the approach advocated here of providing a set of tools and technologies for building customized knowledge management solutions, as opposed to providing a one-size-fits-all knowledge system. Tool integration, however, requires some additional components to the architecture that glue together the functionality of the components. There is also a need for a library for ontology transformations such as filtering and custom inferencing to fit the needs of the various subsystems.
- Building a search engine from an ontology using QuizRDF is a one-click process in contrast to creating an ontology-based presentation with Spectacle, which is a complex programming task. The QuizRDF search engine, however, cannot leverage ontological knowledge that goes beyond the common data model, while the Spectacle presentation can be custom tailored for the automatically extracted ontology (e.g. to present related concepts).
- Ontologies obtained through natural language processing are lightweight ontologies without a solid class hierarchy. This situation will be alleviated by supporting the automated ontology extraction through a repository of background knowledge that contains the domain information not found in the texts and guides the modelling process (e.g. by capturing design decisions such as *what is a class* and *what is an instance*). It is already possible, however, to capitalize on the advantages of automatically extracted ontologies that include, among others, a ranked selection of concepts, cross-taxonomical relationships (relations between concepts) and the automated markup of pages with concepts. This is particularly important for the

Semantic Web where automated approaches and lightweight ontologies will prevail.

- The business case for ontology-based search and navigation is particularly strong for virtual enterprises, such as EnerSearch, whose main value driver is the creation and dissemination of (scientific) knowledge. For this kind of enterprise, the gains from employing ontologies can offset the significant technological risks involved with using advanced semantic technologies.

Virtual organizations provide a fertile test ground to validate the ideas that underlie knowledge management through Semantic Web methods. They have general characteristics that are such that semantic methods promise to be very beneficial. EnerSearch is such a knowledge-intensive virtual organization in which one of the main business ideas is to produce and transfer knowledge from researchers and experts to interested industries. So, internal, and even more importantly, external knowledge management is a key function where semantic methods can prove to be very helpful. But, the proof of the Semantic Web pudding is in the eating.

14

A Future Perspective: Exploiting Peer-to-Peer and the Semantic Web for Knowledge Management

Dieter Fensel, Steffen Staab, Rudi Studer, Frank van Harmelen and John Davies

14.1 Introduction

Over the past few years, we have seen a growing interest in the potential of both peer-to-peer (P2P) computing and the use of more formal approaches to knowledge management, involving the development of ontologies. This penultimate chapter discusses possibilities that both approaches may offer for more effective and efficient knowledge management. In particular, we investigate how the two paradigms may be combined.

This book has presented a set of innovative tools for semantic information processing and thus for more selective, faster, and meaningful user access. This tool environment addresses three areas:

- Acquiring ontologies and linking them with large amounts of data. For reasons of scalability this process must be automated based on information extraction and natural language processing technology. For reasons of quality this process also requires the human in the loop to build and manipulate ontologies using ontology editors.
- Storing and maintaining ontologies and their instances. We developed an

RDF Schema repository that provides database technology and simple forms of reasoning over web information sources.

- Querying and browsing semantically enriched information sources. We developed semantically enriched search engines, browsing and knowledge sharing tools that make use of the machine processable semantics of data provided by the Semantic Web.

The developed technology has been shown to be useful in a number of case studies. It improved information access in the intranet of a large organization (Lau and Sure, 2002). The technology has also been used to facilitate electronic knowledge sharing and re-use for customer relationship management and knowledge management in virtual organizations.

We also encountered a number of shortcomings in our current approach. Ontologies help to establish consensual terminologies that make sense to all users. *Computers* are able to process information based on their machine-processable semantics. *Humans* are able to make sense of this information based on its connection to real-world semantics. However, building up such ontologies that are a pre-requisite for and result of the common understanding of large user groups is no trivial task. A model or 'protocol' for driving the network that maintains the process of *evolving* ontologies is the real challenge for making the *Semantic* Web a reality. Most work on ontologies views them in terms of an isolated theory containing a potentially large number of concepts, relationships, and constraints that further detach formal semantics from them. Conversely, in order to employ their full potential for the Semantic Web, they must be realized as interwoven *networks* that make it possible to deal with the heterogenic needs in the communication processes they are supposed to mediate. It is the *network* of ontologies and their dynamic nature that make future research necessary. So the challenge in current work on ontologies is the quest for what glues ontology networks together in space and time. Instead of a central top-down process we require a distributed process of emerging and aligning ontologies. Proper methodological and tool support for building up networks where the nodes represent small and specialized ontologies is far from well understood – especially not for such a noisy and dynamically changing environment as the web is and will remain. In this chapter, we describe our vision in terms of a set of future steps that need to be taken to bring the results described in earlier chapters to their full potential.

The remainder of this chapter is organized as follows. In Section 14.2, we discuss the main requirements for effective and efficient knowledge management. Section 14.3 takes a detailed look at the requirement on ontologies as a means for supporting communication and information exchange in large, heterogeneous and dynamic networks. Section 14.4 proposes a combination of ontologies and the P2P paradigm as a way to offer much enhanced knowledge management capabilities. Final conclusions are provided in Section 14.5.

14.2 A Vision of Modern Knowledge Management

Knowledge is one of the most crucial success factors for today's enterprises. Therefore, knowledge management (KM) has been identified as a strategically important capability. Clearly, KM is an interdisciplinary task, including human resource management, enterprise organization and cultural considerations as well as IT. However, there is a widespread consensus that IT plays an important role as an enabler for many aspects of a KM solution.

In the past, IT for knowledge management has focused on the management of knowledge containers using text documents as the main repository and source of knowledge. As we have argued in this book, Semantic Web technology, especially ontologies and machine-processable relational meta-data, pave the way to enhanced KM solutions that are based on semantically related knowledge pieces of varying granularity. Although early semantic web-based KM approaches and solutions have shown the benefits of ontologies and related methods, there still exist a large number of open research issues that have to be addressed in order to make semantic web technologies fully effective when applied to KM solutions, as discussed below.

14.2.1 Knowledge Integration

Industrial KM applications need to minimize any kind of development overhead as far as possible. Therefore, a *seamless integration of knowledge creation*, for example, content and meta-data specification, and *knowledge access*, for example, querying or browsing, into the working environment is required (Handschuh and Staab, 2001). Strategies and methods are needed that support the creation of knowledge automatically or semi-automatically as side-effects of activities that are carried out anyway. An example of such an approach is the usage of forms that are generated from an ontology. Since form elements are then linked to the underlying ontology, filling in these forms results in the creation of meta-data which maps directly to the ontology.

14.2.2 Knowledge Categorization

The new field of *emergent semantics* (Mädche et al., 2002) deals with assigning semantic categories and relations to objects by observing user interactions or patterns of user behaviour, and taking into account the user's current context. That is to say, emergent semantics is closely related to ontology learning where the aim is to reduce the overhead of building-up and maintaining ontologies. A P2P knowledge management system is ideally suited to take advantage of emergent semantics, because many typical knowledge worker actions have implicit semantics attached to them, such as:

- sending (new) email messages with subject descriptions;
- organizing documents into folders;
- clustering knowledge (e.g. presenting overlapping hierarchies of bookmarks);
- developing schemata (XML DTDs, XML schemata, RDF schemata, etc.).

14.2.3 Context Awareness

Access to, as well as presentation of, knowledge has to be context-dependent. Since the context is set up by the current business task and thus by the business process being handled, a tight integration of business process management and knowledge management is required (Abecker et al., 2000). KM approaches which can manage knowledge pieces provide a promising starting point for smart push services that will proactively deliver relevant knowledge for carrying out the task at hand. For example, the Smart Task approach (Staab and Schnurr, 2000) uses knowledge of the sub-task being carried out as well as the content of documents being handled for submitting dynamically defined queries to the knowledge warehouse. The delivered answers provide knowledge that is of importance for the given context, that is, the performed sub-task and the processed documents. Such a context-dependent knowledge delivery is a promising approach for avoiding information overload for the knowledge worker. *Location-aware services* are a specialization of this technique. They provide the knowledge worker with the most relevant knowledge pieces depending on where he or she is situated, for example, in the basement of a construction site or at a trade fair (Klemke and Nick, 2001).

14.2.4 Personalization

Contextualization has to be supplemented by personalization. Taking into account the experience of the user and his or her personal needs is a prerequisite, on the one hand, for avoiding information overload and, on the other hand, for delivering knowledge at the right level of granularity. Again, Semantic Web technology provides a promising basis since ontologies and associated meta-data deliver semantic information about the user behaviour and user interests.

Personalization may be supported by technologies such as statistics and data mining. However, future knowledge management solutions will rely on more sophisticated applications of data mining that are currently heavily under-explored: document repositories will be mined for interesting relations, integrating data from the enterprise value chain. For instance, customer response should be mined in conjunction with production data in order to find faults in single production steps.

14.2.5 Knowledge Portal Construction

The development of knowledge portals serving the needs of companies or communities is still a more or less manual process. Ontologies and related meta-data provide a promising conceptual basis for generating (parts of) such knowledge portals (Jin et al., 2001). Obviously, conceptual models of the domain, the users and the tasks are needed among others. Generation of knowledge portals has to be supplemented with the (semi-)automatic evolution of portals (Staab and Mädche, 2001; Hotho et al., 2001). Since business environments and strategies change rather rapidly, KM portals have to be kept up-to-date in this fast changing environment. Evolution of portals also includes the aspect of 'forgetting' outdated knowledge.

KM solutions will be based on a combination of intranet-based functionalities and mobile functionalities in the very near future. Semantic Web technologies are a promising approach to meet the needs of the mobile environments, such as location-aware personalization and adaptation of information presentation to the specific capabilities of mobile devices, that is, the presentation of the required information at an appropriate level of granularity. In essence, employees should have access to the KM application *anywhere* and *anytime*. However, in order to be able to deal with large amounts of knowledge, it must be served at different levels of granularities. For instance, a search on the desktop might also pop up related links and many details, but on the palmtop one must restrict knowledge pieces to the bare minimum.

14.2.6 Communities of Practice

The importance of knowledge sharing and re-use in knowledge management in order to share best practice and prevent duplication of effort has led to much interest in *communities of practice*. Recent ethnographic studies of workplace practices indicate that the ways people actually work often differ fundamentally from the ways organizations describe that work in manuals, organizational charts and job descriptions. The term 'community of practice' describes the informal groups where much knowledge sharing takes place and has been increasingly applied in the knowledge management context. They are typically not a formal team but an informal network, each sharing in part a common agenda and shared interests. In one example, it was found that a lot of knowledge sharing among photocopier engineers took place through informal exchanges, often around the coffee point. As well as such local, geographically based communities, trends towards flexible working and globalization have led to interest in supporting global communities using Internet technology. The challenge for organizations is to support such communities and make them effective. Semantic Web technology can be used to support communities of practice more effectively: ontologies, as consensual domain models directly address a key concern of such communities – how to represent and reason about a shared world model or conceptualization. In Chapter 10, we saw an early example of the use of Semantic Web technology to support communities of practice.

14.2.7 P2P Computing and its Implications for KM

P2P computing combined with Semantic Web technology will be an interesting and effective alternative to the more centralized KM solutions that are currently implied by ontology-based solutions. The term 'peer-to-peer' became commonplace in IT towards the second half of 2000. The application largely responsible was Napster, the music file-sharing application. Many similar P2P applications appeared at around the same time, such as Gnutella, Kazaa, and Freenet. P2P applications effectively turn an individual user's computer from primarily a display device (showing emails, documents and WWW pages) into an active Internet-based computing resource. The success of these initial applications has led to interest in the exploitation of P2P architectures in business applications. Leveraging the spare CPU cycles and storage capacity of an organization's PCs will, it is anticipated, bring greater flexibility and a better return from a company's IT investments. In short, P2P allows a highly decentralized computer network architecture, with computers (and hence applications) able to interact with one another with little or no centralized control.

P2P scenarios open up the way to derive consensual conceptualizations among employees within an enterprise in a bottom-up manner. Thereby, another requirement may be fulfilled: Future knowledge management solutions should be nearly free of administration in order that they may be used by everyone including private persons and small cooperating companies perhaps as part of a virtual company. We discuss further the relationship between P2P technology and KM in Section 14.4.

14.2.8 Virtual Organizations and their Impact

Virtual organizations are becoming increasingly important in business scenarios characterized by decentralization and globalization. Obviously, semantic interoperability between different knowledge sources as well as trust is a must in inter-organizational KM applications. The process of setting up a virtual organization may also take profit of semantic descriptions of competencies and experiences of prospective partners. This enables a semantic matching process selecting the most appropriate collection of partners from a group of potentially interesting partners.

14.2.9 eLearning Systems

The smooth integration of KM applications, for example, skill management, with eLearning is an important field enabling a lot of synergy between these two areas. KM solutions and eLearning have to be integrated from both an organizational and an IT point of view. KM solutions that have an explicit notion of context and user profile provide an excellent environment for tailoring the learning situation to the specific needs of the learner. Clearly, interoperability and/or integration of (meta-data) standards of the eLearning community, for example, LOM or Ariadne, with the Semantic Web meta-data standards are needed to realize such an integration (Maurer and Sapper, 2001; Stojanovic et al., 2001). Attaching semantic meta-data to learning modules enables the context specific configuration of learning modules and their adaptation to the specific user needs. Of course, the ontologies being used to specify the learning module meta-data have to be aligned with the ontologies defining the context and user profile.

14.2.10 The Knowledge Grid

Further application areas of KM and the Semantic Web will be the Knowledge Grid, particularly with regard to eScience applications. In subjects such as genomics, astrophysics, bioengineering, or biochemistry there are massive data sources, however little access to these data by semantic means. Thus, these data sources are plagued by the same effects as knowledge in the WWW. They are hardly digestible by software and humans alike and need an underlying infrastructure such that people may better share experiences, analyses may be easily redone, and data may be easily integrated. For instance, knowledge portals may be used to assist analysts in selecting, combining and aggregating knowledge in the life sciences area. Since life sciences evolve very quickly and so does the terminology of the field, a further important aspect is the association of new words that show up in new documents with the appropriate concepts from the respective ontologies.

14.2.11 Intellectual Capital Valuation

In order to determine the outcome of knowledge management or eLearning undertakings, knowledge should be accounted for in the same way as other goods. An important part of *knowledge valuation* will be the structuring of the underlying knowledge based on which the valuation proper may be done.

14.3 A Vision of Ontologies: Dynamic Networks of Meaning

As argued before, ontologies glue together two essential aspects that help to bring the web to its full potential:

- Ontologies define *formal* semantics for information, consequently allowing information processing by a computer.
- Ontologies define *real-world* semantics, which makes it possible to link machine processable content with meaning for humans based on consensual terminologies.

Especially the latter aspect is still far from being studied to its full extent: how can ontologies be used to communicate real-world semantics between human and artificial agents? In answering this question we will argue that most current work on ontologies is mis-focused in part, in that it ignores the main problems in building and using them. Most of the work on ontologies views them in terms of an isolated theory that covers a potentially large number of concepts, relationships, and constraints that further attach formal semantics from them. Here, we take a much broader view of ontologies: *Ontologies must have a network architecture and ontologies must be dynamic.* That is, ontologies have to deal with:

- *Heterogeneity in space: ontology as networks of meaning.* From the very beginning, heterogeneity has been an essential requirement for this ontology network. Tools for dealing with conflicting definitions and strong support in interweaving local theories are essential in order to make this technology workable and scalable, just as the first generation web was workable and scalable because it was error-tolerant with regard to hyperlinking.
- *Development in time: living ontologies.* Originally, an ontology should reflect the 'truth' of a certain aspect of reality. It was the holy task of a philosopher to find such truth. Today, ontologies are used as a means of

exchanging meaning between different agents. They can only provide this if they reflect an inter-subjectual consensus. By definition, they can only be the result of a social process. For this reason, ontologies cannot be understood as a static model. An ontology is as much required for the exchange of meaning as the exchange of meaning may influence and modify an ontology. Consequently, *evolving* ontologies describe a process rather than a static model. Having protocols for the process of evolving ontologies is the real challenge. Evolving over time is an essential requirement for useful ontologies. As daily practice constantly changes, ontologies that mediate the information needs of these processes must have strong support in *versioning* and must be accompanied by *process models* that help to organize evolving consensus.

In the following sub-section, we take a philosophical view of ontologies, helping to clarify their very nature. Then, we discuss in more detail the aspects of heterogenity in space and time. We conclude with a short summary.

14.3.1 Ontologies or How to Escape a Paradox

Every first-year philosophy student has heard about the vicious circle in trying to explain our ability to communicate as a means of exchanging meaning and creating understanding between human beings. On the one hand, people can only communicate and exchange meaning based on a common understanding of symbols and intentions. Thus, an agreed set of symbols and a consensual interpretation thereof is a prerequisite for communication. On the other hand, such a set of symbols and a consensual interpretation can only be established as a result of communication.[1] In other words, the result of successful communication is, at the same time, a prerequisite for it. Consequently, its existence is required for explaining its existence. Our first-year philosophy student may also have learned how to overcome such a paradoxical situation. There must be an underlying *process* that takes both sides as intermediate and frequently repeated sub-steps that rely on something that mediates between the extremes (Hegel, 1812). This would make successful communication and a shared set of understandings merely two sides of the same coin. The reader may have found our arguments 'too philosophical'. However, we want to underline the principal difference between viewing ontologies as '*true' models of the real world* or as *steps in a process of organizing evolving consensus*. For this reason, a brief argument on the cyclic nature of understanding and communication would seem appropriate. From an abstract philosophical perspective, it seems like a miracle that any two human beings can understand each other. Meaning and

---

[1] At least as long as this interpretation is not hard-coded via instincts.

intention cannot be exchanged or expressed directly. Nor can we access the actual meaning that is perceived and understood by our counterpart. We can only express our intention by some action that influences the perception of our counterpart. And we can only guess what this is supposed to mean to him by analysing his behaviour as much as it is reflected in our perception (Hegel, 1812). Thus, establishing meaning and communication (to exchange meaning) is a *process* by definition. People can only establish shared meaning and communicate it to each other through a process in which they coordinate some of their actions to achieve common goals. Therefore, from the very beginning[2] it can only be a social process that creates a shared understanding, which serves as a basis for exchanging meaning with communicative symbols. Following this argument, it becomes fairly rather clear that *the* ontology to which everybody subscribes does not exist. Instead, ontologies arise as pre-requisites and results of cooperation in certain areas reflecting task, domain and sociological boundaries. The web weaves billions of people together to support them in their information needs. Similarly, ontologies can only be viewed as a network of interwoven ontologies. This network of ontologies may have overlapping and excluding pieces, and must be as dynamic in nature as the dynamics of the process it underlies. This view on *ontologies as dynamic networks of formally represented meaning* is what we want to stress here. Most of the work on ontologies views them in terms of an isolated theory that covers a potentially large number of concepts, relationships, and constraints that further detach formal semantics from them. Here, we take a much broader view of ontologies. To recap, there are two main dimensions in which these mediators of communication differ from current work on ontol-ogies: *ontologies must have a network architecture and ontologies must be dynamic*.

14.3.2 Heterogeneity in Space: Ontology as Networks of Meaning

An island of meaning must be interwoven to form more complex structures enabling exchange of information beyond domain, task and sociological boundaries. This has two practical implications. Firstly, tool support must be provided to define local domain models that express a commitment of a group of agents that share a certain domain and task and that can agree on a joined world view for this purpose. A great deal of work has already been done in this area and significant methodological support is available (for a survey, see Fensel et al., 2002b). Secondly, these local models must be interwoven

---

[2] In principle, it is not even important whether another agent actually thinks. He 'understands' our communicative acts properly if they are properly contained within the framework of his approach to cooperation with us.

with other models, such as the social practice of the agents that use ontologies to facilitate their communicational needs. Little work has been done in this latter area.[3] We no longer talk about a single ontology, but rather about a network of ontologies. Links must be defined between these ontologies and this network must allow overlapping ontologies with conflicting – and even contradictory – conceptualizations. From the very beginning, heterogeneity has been an essential requirement for this ontology network. Tools for dealing with conflicting definitions and strong support in interweaving local theories are essential in order to make this technology workable and scalable.[4]

Gnutella, a P2P network, is a case in point (Oram, 2001). Agents were able to enter and leave the network dynamically. They could also communicate with a local environment of other agents. This network was dynamically set up and collapsed according to the joined needs of a group of agents. Current work on ontologies that focuses either on local domain theories or on principles, structures and content of the right 'upper-layer' (general purpose) ontology are far from supporting such a vision. What is needed is a focus on:

- *linking local conceptualizations* that deal with heterogeneous definitions and personalized views;
- support in easy *configuration and re-configuration of such networks* according to the communication needs of agent coalitions; and
- methods and tools that help agents *to organize consensus*, allowing them to exchange meaning.

Ontologies ensure communication between various agents. They are 'right' if they fulfil this purpose.

14.3.3 Development in Time: Living Ontologies

Originally, an ontology should reflect the 'truth' of a certain aspect of reality. It was the holy task of a philosopher to find such truth. Today, ontologies are used as a means of exchanging meaning between different agents. They can only provide this if they reflect an inter-subjectual consensus. By definition, they can only be the result of a social process. This gives ontologies a dual status for the exchange of meaning:

- Ontologies as pre-requisite for consensus. Agents can only exchange meaning when they have already agreed on a shared body of meaning reflecting a consensual point of view on the world.

---

[3] An experiment for developing a shared ontology of a world-wide distributed research community is described in Benjamins et al. (1999).

[4] Steps in the direction we recommend are supported in ontology mappings (cf. Omelayenko and Fensel, 2002) and re-use (cf. Ding and Fensel, 2002).

- Ontologies as a result of consensus. Ontologies as consensual models of meaning can only arise as result of a process where agents agree on a certain model of the world and its interpretation.

Thus, ontologies are as much a pre-requisite for consensus and information sharing as they are the results of them. For this reason, ontologies cannot be understood as a static model. An ontology is as much required for the exchange of meaning as the exchange of meaning may influence and modify an ontology. Consequently, *evolving* ontologies describe a process rather than a static model. Having protocols for the process of evolving ontologies is the real challenge. Evolving over time is an essential requirement for useful ontologies. As daily practice constantly changes, ontologies that mediate the information needs of these processes must have strong support in *versioning*[5] and must be accompanied by *process models* that help to organize consensus.

Centralized process models have standardization bodies, such as the central clearing unit. This central unit may soon pose problems to the scalability of the entire process. Often, such standardization is slow and leads to mongrelized results. Decentralized process models for consensus achievement can be based on the natural consensus of working networks. They can, therefore, reflect a consensus that is true, has proven useful and is used widely. In this context, one may want to take a look at P2P, where networks arise and are configured dynamically according to the shared interests of loosely coupled groups.

14.4 Peer-2-Peer, Ontologies and Knowledge

In today's knowledge-based economy, the competitiveness of enterprises and the quality of work life are directly tied to the ability to effectively create and share knowledge both within and across organizations. In this section, we discuss solutions for knowledge management based on integrating the Peer-2-Peer paradigm with ontology-based information access, that is, Semantic Web technology (also cf. Nejdl et al., 2002). First, we discuss the shortcoming of both approaches when realized in isolation. Second, we define challenges in combining them and finally we sketch major advantages of such a unifying approach.

14.4.1 Shortcomings of Peer-2-Peer and Ontologies as Isolated Paradigms

P2P computing (in its various flavours) is currently a trend in intranet and Internet applications. Analysts like the Gartner Group expect its fast growing

---

[5] Klein and Fensel (2001) provide an analysis for a versioning framework.

importance for knowledge management solution in large companies and organizations. Companies like Microsoft, IBM, Intel, SAP (together with Groove
Networks), and Sun (developing JXTA) have announced first solutions in this
area. Emerging P2P solutions are particularly well suited to the increasingly
decentralized nature of today's organizations, be it a single enterprise or a
dynamic network of organizations. They make it possible for different participants (organizations, individuals or departments within an organization) to
maintain different views of the world while exchanging information. They
also circumvent the bottlenecks associated with more traditional solutions,
which rely on one or a small number of centralized servers. At the same
time, because they rely on keyword search and rather simple knowledge
representation techniques, today's peer-to-peer solutions are extremely
limited. They cannot easily support the introduction of new concepts, they
make it difficult to determine whether two terms are equivalent, and generally
can only support very limited levels of automation – all types of functionality
which Semantic Web technologies have been shown to support.

The current state-of-the-art in knowledge management solutions still
focuses on one or a relatively small number of highly centralized knowledge
repositories with *ontologies* as the conceptual backbone for knowledge
brokering. As it turns out, this assumption is very restrictive, because

- it creates major bottlenecks and entails significant administrative over-
  heads, especially when it comes to scaling up to large and complex
  problems;
- it does not lend itself to easy maintenance and the dynamic updates often
  required to reflect changing user needs, dynamic enterprise processes or
  new market conditions.

In contrast *P2P computing* offers the promise of removing many of these
limitations. The essence of P2P is that nodes in the network directly exploit
resources present at other nodes of the network without intervention of any
central server. The tremendous success of networks like Napster and Gnutella,
and of highly visible industry initiatives such as Sun's JXTA, as well as the
Peer-to-Peer Working Group including HP, IBM and Intel, have shown that
the P2P paradigm is a particularly powerful one when it comes to sharing files
over the Internet without any central repository, without centralized administration, and with file delivery dedicated solely to user needs in a robust,
scalable manner. At the same time, today's P2P solutions support only limited
update, search and retrieval functionality. P2P solutions currently have the
following limitations:

- In many projects, P2P is discussed as a solution at the protocol level (no client–server dichotomy) and as a means for distributing disk space. However, this is of minor importance for improved service in knowledge management. Here it is the actual sharing of information and knowledge which needs to be supported and not the organization of disk storage or network traffic.

- Existing solutions such as Napster or Gnutella provide limited support in information and knowledge sharing. Napster supports only keyword-based search of music titles and author names, Gnutella does not provide any pre-defined search support. Each Gnutella client is completely free in how it interprets a query.
- Leading industry efforts such as JXTA by Sun Microsystems are limiting P2P services to string matching. No support for shared ontologies is provided. Queries are specified in arbitrary XML formats. No use is being made of the opportunities to use RDF/RDF Schema for expressing shared vocabularies. Finally, JXTA limits query-answering to using resources in a single location, while in fact many queries will require the combination of information retrieved from different sources at different locations.
- Peer selection is currently not really based on content. This needs to be improved to route a query to the next knowledgeable peer rather than some arbitrary one.

These flaws make current P2P systems unsuitable for knowledge sharing purposes. Key to the success of combining P2P solutions with Semantic Web technologies is the use of emergent semantics. Emergent semantics can build on lightweight and/or heavyweight ontologies that different individuals, departments or organizations have created. It considers the overlap between ontology definitions and the use of concepts and relations with actual data in order to extract shared ontologies for sets of individuals or groups of people. Intelligent tools will use such definitions to ensure that knowledge will be appropriately structured, so that it can be easily found. Knowledge management can occur in a distributed fashion without overhead through central administration.

14.4.2 Challenges in Integrating Peer-2-Peer and Ontologies

We aim for a P2P based knowledge management approach that integrates the advantages of Semantic Web-based knowledge management technology developed in successful IST projects like On-To-Knowledge, KnowNet, or Comma. Figure 14.1 illustrates the intended advantages. It depicts a qualitative comparison of benefits (time saved or redundant work avoided expressed

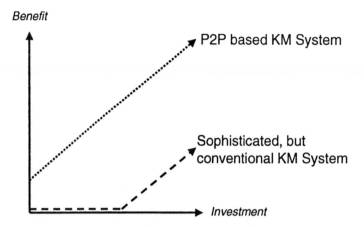

Figure 14.1   Qualitative comparison of benefits resulting from investments in KM systems

financially) gained by using a KM system against the amount of knowledge in that system. Of course, explicitly modelled ontologies may increase the benefits brought by *any* knowledge management solution, because they may improve the accuracy of knowledge access and sharing. The solutions we describe, however, may produce benefits even with near zero investment – in contrast to conventional knowledge management systems that need an extensive and expensive set-up phase.[6] Conventional knowledge management repositories will still appear as just another, powerful peer in the network. Hence, a combined Semantic Web and P2P solution may always outperform the sophisticated, but conventional centralized system.

Challenges for such an approach are:

- *Peer selection service*. In order to receive the right answers without flooding the peer network with queries one must ask the 'right' peers. Ontology-based peer selection mechanisms need to exploit similarity of ontologies for this purpose.
- *Variation of ontologies*. Different peers will use different, though overlapping ontologies. *Alignment, mapping* and *visualization tools* will have to cope with different ontologies, even though no alignments are explicitly specified. Some of the alignments and the mappings may be found by analysis of peer knowledge using methods of the just emerging field of

---

[6] We are focussing here on the investment required in IT systems. There may or may not also be a cost with regard to cultural change programmes and so forth.

emergent semantics (e.g. same file categorized to different concepts indicates alignment).

- *Lack of ontological precision.* Ontologies will be produced from various user interactions, like classifications into folders or usage of meta-data. Ontology definitions will be imprecise and 'sloppy' ontologies will be the norm rather than the exception. An inference engine for these ontologies must be able to ask and answer queries to peers in a robust, scalable, often locally contained, manner.

- *Ontological drift.* In a P2P environment, one cannot expect *any* maintenance of the ontologies to happen (in fact, users will often not know what is in the ontologies on their machine). As a result, we must design mechanisms that allow the ontologies to update themselves, in order to cope with ontological drift. Based on the queries and answers elsewhere in the P2P network, ontologies will have to adjust their own definitions accordingly.

Each of these challenges will require significant research in ontology engineering technology.

14.4.2.1 Peer Selection Service

Clearly, the success of a P2P network relies on the quality of peers that each member of the network selects.

Example: Investment Analysis. *If an investment analyst in the TIME[7] market performs a domain-specific query, it is highly improbable that a colleague of his/hers who specializes in the Life Science Domain will know the answer. Correspondingly, the query should rather be routed to colleagues who work in similar domains in order to receive many useful answers.*

Selection of peers in existing P2P networks is typically based on network related parameters, such as ping-time, number of network-hops, etc. Clearly, in an ontology based P2P service, one would rather want to select peers on the basis of their conceptual proximity to one's own ontology – possibly including contextual information (e.g. what activity the user is currently engaged in).

Continuing the example from above, we would expect that colleagues with overlapping tasks would also have overlapping ontologies that describe their view of the domain.

Therefore, routing of queries should depend on some kind of 'similarity measures' between ontologies, which can be used to select peers with similar ontologies, that is, peers that are likely to have answers to my queries or peers

---

[7] TIME stands for telecommunication, information technology, multimedia and e-business.

that are likely to even understand my queries. So far, no robust notion of 'conceptual distance' or 'semantic similarity' between ontologies exists in the current literature; it needs to be developed and evaluated.

14.4.2.2 Dealing with Variation of Ontologies

Different peers will use different, though overlapping ontologies. *Alignment, mapping* and *visualization tools* will have to cope with different ontologies, even though no alignments are explicitly specified.

Example: Virtual Enterprise Case Study. *Two SMEs, let us call them design.com and producing.com, have agreed about cooperation and about sharing knowledge resources in order to watch current and upcoming technology in their markets. Therefore, they both collect product announcements, but view them from different angles. design.com must know which possibilities they are given by new technology (what), while producing.com must know how to handle the same new technology (how). The what and the how refer to different views of concepts in the two organizations.*

Some of the alignments and the mappings may be found by analysis of peer knowledge using methods of the just emerging field of *emergent semantics* (e.g. same file categorized to different concepts indicates alignment).

The classification of the same announcement under the two different concepts at the two companies may be exploited for proposing a more general alignment between design.com's what and produce.com's how. Furthermore, one may automatically translate queries for design.com's what into produce.-com's how, allowing the user to navigate the two peers and their ontologies while being visually reminded of the underlying differences and commonalities.

14.4.2.3 Robust Inferencing

Current ontology-based applications rely on careful engineering of the terms in the ontology. Terms are carefully defined to obtain maximally clean and useful relationships between them (proper subset relations, complement relations, strict domain/range restrictions, etc). All of the current inference techniques for ontologies (querying, inferencing, consistency checking, subsumption checking, etc.) rely on such clean relationships. The same holds for current applications based on multiple ontologies. Multiple ontologies can currently only be integrated through careful, handcrafted engineering.

In a P2P setting, ontologies will have to be obtained with no or very little human intervention (e.g. by 'scraping' ontologies from existing resource structures such as file hierarchies, mail folders, etc.). These ontologies will therefore not be of the same high quality as current human-engineered ontologies, and many of the relationships in an ontology will be incomplete or even incorrect. For instance, sub-class definitions will not be precise or instances will sometimes be miss-classified. As a result, many (if not all) of the existing inference techniques fail to produce useful results on such 'sloppy' ontologies. Even a single miss-classified instance is sufficient to invalidate a sub-class relationship, and even one wrong value for a single attribute is sufficient to make a class inconsistent.

Example: Investment Analysis.   *An investment analyst in the TIME market may erroneously mis-classify information about the Danish company 'Excel Data' into the category 'Spreadsheet Software'. Complementary knowledge may describe that 'Excel Data' is a company and, therefore, the logical theory becomes inconsistent – leading in the 'logical worst case' to arbitrary answers to any question, for example, that 'Bill Gates' is a product.*

What is needed in a P2P setting are more robust forms of reasoning that can cope with limited amounts of incompleteness and inconsistencies. Such robust inference techniques will be able to identify when a sub-class relationship 'almost holds' (even though a few instances may be mis-classified), or when an ontology is 'almost consistent' (by ignoring a few local inconsistencies).

Continuing our example, the mis-classification of 'Excel Data' and corresponding inconsistencies must be 'contained' in small niches. Then wrong answers about 'Excel Data' will be given, but 'Bill Gates' will still be known as founder of Microsoft.

Existing work in the field of knowledge representation on approximate inference (e.g. Cadoli and Schaerf, 1995; Dalal, 1996) and local containment of inconsistencies (Wasserman, 2000) is expected to provide the required framework to tackle the problem of reasoning with sloppy ontologies.

14.4.2.4 Handling Ontological Drift

Ontological drift is a natural phenomenon that appears in almost every domain.

Example: Investment Analysis.   *Notions about the capabilities of software and hardware have changed tremendously in the TIME market within a few years. 'eCommerce' first almost exclusively referred to business-to-consumer* *sales. Over time it changed its meaning and now also includes business-to-business, customer-to-customer, business-to-employee, etc. interactions.*

Such ontological drift means that ontologies must be constantly maintained. We must regularly update the definitions of terms and relations in an ontology. Such maintenance is possible in current applications, because they all employ a single ontology in a centralized location.

In a P2P environment, one cannot expect *any* maintenance to happen on the ontologies (in fact, users will often not know what is in the ontologies on their machine, let alone that they can perform maintenance on them). As a result, we must design mechanisms that allow the ontologies to update themselves, in order to cope with ontological drift. Based on the queries and answers elsewhere in the P2P network, ontologies will have to adjust their own definitions accordingly: if other ontologies give different answers to queries, possibly update your own definitions; possibly based on 'confidence' in the other ontologies; do this on a sliding scale of truth values of the definitions (instead of simply binary true/false); use metaphors from social science (opinion-forming, rumour-spreading, etc.).

14.5 Conclusions

The combination of ontologies and the P2P paradigm may provide promising directions for future research and tool development and we discuss some of these possibilities below.

14.5.1 P2P for Knowledge Management

We propose to develop and validate P2P-based knowledge management solutions, where knowledge is updated and searched in a completely decentralized manner, thereby moving away from the limitations of today's knowledge management solutions, which are organized around one or a small number of centralized knowledge clusters. For this purpose we must resolve the issues around peer selection services based on knowledge structure and content (cf., e.g. peer selection services).

14.5.2 P2P for Ontologies

Ontology technology should be moved to the next level by letting the semantics emerge from the way in which knowledge is accessed and used. Centralized ontologies are too static when they are defined in a slow, centralized process while the users' interests/perspectives they reflect change fast and frequently. We propose to develop ontology technology that gives participants the freedom to use their own ontology structures. The P2P knowledge management system will be able to take advantage of differences (cf. 'ontological drift' from above as one example). It will dynamically come up with ontologies that constitute common denominators for groups of participants as well as with proposals for mappings between ontologies (e.g. through common names of documents at different nodes in the network). Thus, one may come to semantic conclusions without centralized ontologies.

14.5.3 Ontologies for P2P and Knowledge Management

With P2P-based solutions, existing ontology-based knowledge management approaches do not stay the same. Ontologies give participants the semantic accuracy for finding the answers to their problems, but ontology use must change; for example, ontology inferencing must be adapted (cf. 'robust inference' above) or new visualization techniques must be investigated in order to deal with the distinction between local and shared ontologies. The use of emergent semantics gives all the freedom to the participant that he/she wants, but ontology tools must be adapted to meet the challenges of the new paradigm.

14.5.4 Community Building

Because communities are not defined by 'what repository you are using', but rather by 'what knowledge is important to you', people have new possibilities to form communities taking full advantage of P2P communication.

Over the past few years, we have seen a growing interest in the potential of P2P computing and the use of more formal approaches to knowledge management, involving the development of ontologies. At the same time, to date, nothing has been done to combine the power of these approaches.

We began by discussing some of the main requirements for effective and efficient knowledge management. We then examined the requirement on ontologies as a means for supporting communication and information exchange in large, heterogeneous and dynamic networks. We proposed a combination of ontologies and the P2P paradigm as a way to offer much enhanced knowledge management capabilities.

Our argument has been that only by bringing together Semantic Web (specifically, ontologies) and P2P technology can we fully realize their potential to enhance dramatically the capabilities offered by today's knowledge management solutions.

15

Conclusions: Ontology-driven Knowledge Management – Towards the Semantic Web?

John Davies, Dieter Fensel and Frank van Harmelen

The global economy is rapidly becoming more and more knowledge intensive. Knowledge is now widely recognized as the fourth production factor, on an equal footing with the traditional production factors of labour, capital and materials. Managing knowledge is as important as the traditional management of labour, capital and materials.

In this book, we have shown how Semantic Web technology can make an important contribution to knowledge management. A prerequisite to applying many of the technologies described in this volume is that knowledge is sufficiently computer-accessible. This coincides with the vision of the Semantic Web at large: making knowledge machine-accessible rather than limiting the role of the computer to displaying knowledge sources on a screen for interpretation by humans. In Chapter 2, we discussed the representation languages that are required for making knowledge computer-accessible in a way that is both expressive, logically well-founded, and compatible with current IT standards such as XML.

In any knowledge management initiative, technology alone is not sufficient: knowledge management is a sociotechnical discipline. Chapter 3 described a methodology that should be used in order to effectively apply the technology from this volume.

Chapters 5–11 described a wide range of technologies to exploit the Semantic Web for knowledge management: an ontology editor to capture human expertise, an environment for collaborative knowledge sharing, tools for automated concept extraction from knowledge sources, support for semantic search and navigation through knowledge sources, among others.

Chapters 12 and 13 showed how these tools, under the guidance of the methodology, have been used in realistic case studies in very different companies, and under widely varying conditions.

All this demonstrates that Semantic Web technology can indeed be used for knowledge management. This technology was deployed on a company wide-scale, with company intranets as the technological base. But the title of this book also hints at something else: it contains the promise that the knowledge management technologies of this book would also be steps towards a *worldwide* Semantic Web, rather than being limited to company-wide intranets, highly valuable though such applications are.

We do indeed believe that this is the case: much of the same technology that was developed in this book for the purposes of company-wide knowledge management has also been shown to be useful on a much larger scale. A follow-up language of OIL is currently in large-scale use on the web (http://www.daml.org), tools like OntoEdit are being used to mark up pages on the web, infrastructure technology such as Sesame is now in use for web applications that are not just limited to internal company use.

That is not to say, however, that the results from this book are now ready to be applied in an industrial setting in a routine fashion. The tools are in the main prototypes, the methodology needs to mature, the case studies have been relatively costly, labour intensive and have required input from skilled specialists. Much exciting work remains to be done to make the transition from the pages of this book into real practice, in real industry, on the real web.

References

Abecker, A., Bernardi, A., Maus, H., Sintek, M. and Wenzel, C. (2000) Information supply for business processes: coupling workflow with document analysis and information retrieval. Knowledge Based Systems, 13(5): 271–284.

Ackerman, M.S., McDonald, D., Lutters, W. and Marumatsu, J. (1999) Recommenders for expertise management. In Workshop on Recommender Systems: Algorithms and Evaluation, SIGIR '99.

Ahmed, K. and Benbrahim, M. (1995) Text summarisation: the role of lexical cohesion analysis. New Review of Document and Text Management, 1: 321–335.

Alexaki, S., Christophides, V., Karvounarakis, G., Plexousakis, D. and Tolle, K. (2000) The RDF Suite: Managing Voluminous RDF Description Bases. Technical report, Institute of Computer Science, FORTH, Heraklion, Greece.
See http://www.ics.forth.gr/proj/isst/rdf/rssdb/rdfsuite.pdf.

Ankolenkar, A., Burstein, M., Hobbs, J.R., Lassila, O., Martin, D.L., McDermott, D., McIlraith, S.A., Narayanan, S., Paolucci, M., Payne, T.R. and Sycara, K. (2002), DAML-S: web service description for the semantic web, In The First International Semantic Web Conference (ISWC), June.

Arprez, J.C., Corcho, O., Fernandez-Lopez, M. and Gomez-Perez, A. (2001) WebODE: a scalable workbench for ontological engineering. In Proceedings of the First International Conference on Knowledge Capture (K-CAP) October 21–23, Victoria, BC, Canada.

Auer, T. (2000) Wissens-Sicherung im Kontext mit der Personal Entwicklung. Schweizer Arbeitgeber, 10.

Banerjee, J., Kim, W., Kim, H.-J. and Korth, H.F. (1987) Semantics and implementation of schema evolution in object-oriented databases. SIGMOD Record. In Proceedings of the Conference on Management of Data, Vol. 16(3), pp. 311–322.

Bates, M.J. (1986) An exploratory paradigm for online information retrieval. In Brookes, B.C. (ed.), Intelligent Information Systems for the Information Society. Amsterdam: North-Holland.

Becerra-Fernandez, I. (2000) Facilitating the online search of experts at NASA using expert seeker people-finder. In Reimer, U. (ed.), 3rd International Conference on Practical Aspects of Knowledge Management PAKM 2000.

Bechhofer, S., Horrocks, I., Patel-Schneider, P.F. and Tessaris, S. (1999) A proposal for a description logic interface. In Lambrix, P., Borgida, A., Lenzerini, M., Mueller, R. and Patel-Schneider, P. (eds.), Proceedings of the International Workshop DL'99, July 30–August 1, Linköping, Sweden.

Bechhofer, S., Goble, C. and Horrocks, I. (2001a) DAML+OIL is not enough. In Proceedings of the First Semantic Web Working Symposium (SWWS'01) CEUR Electronic Workshop Proceedings, pp. 151–159. http://ceur-ws.org/

Bechhofer, S., Horrocks, I., Goble, C. and Stevens, R. (2001b) OilEd: a reasonable ontology editor for the semantic web. In Proceedings of the Joint German/ Austrian Conference on Artificial Intelligence (KI 2001), Lecture Notes in Artificial Intelligence, no. 2174, pp. 396–408. Berlin: Springer-Verlag.

Beckett, D. (2001) The design and implementation of the Redland RDF application framework. In Proceedings of Semantic Web Workshop of the 10th International World Wide Web Conference, May, Hong-Kong, China.

Benatallah, B. and Tari, Z. (1998) Dealing with version pertinence to design an efficient schema evolution framework. In Proceedings of an International Database Engineering and Application Symposium (IDEAS'98), pp.24–33, July 8–10, Cardiff, UK.

Benjamins, R., Fensel, D., Decker, S. and Gomez Perez, A. (1999) (KA)^2: building ontologies for the internet: a mid-term report, International Journal of Human-Computer Studies, 51(3): 687–712.

Bernaras, A. Laresgoiti, I. and Corera, J. (1996) Building and reusing ontologies for electrical network applications. In Proceedings of the European Conference on Artificial Intelligence ECAI-96.

Berners-Lee, T., Hendler, J. and Lassila, O. (2001) The semantic web. Scientific American, May.

Booch, G,. Rumbaugh, J. and Jacobson, I. (1998) The Unified Modelling Language User Guide. Reading, MA: Addison-Wesley.

Box, D., Ehnebuske, D., Kakivaya, G., Layman, A., Mendelsohn, N., Nielsen, H.F., Thatte, S. and Winer, D. (2000) Simple object access protocol (SOAP) 1.1. W3C note, World Wide Web Consortium, May 2000. See http://www.w3.org/tr/soap/.

Bremdal, B.A. (2000) Summarizer. CognIT a.s, white paper.

Bremdal, B.A. and Johansen, F. (2000) CORPORUM technology and applications. CognIT a.s, white paper.

Bremdal, B.A., Johansen, F., Spaggiari, Ch., Engels, R. and Jones, R. (1999) Creating a learning organisation through content based document management, CognIT a.s white paper.

Brickley, D. and Guha, R.V. (2000) Resource Description Framework (RDF) Schema Specification 1.0. Candidate recommendation, World Wide Web Consortium, March. See http://www.w3.org/tr/2000/cr-rdf-schema-20000327.

Broekstra, J. and Kampman, A. (2001a) Query language definition. On-To-Knowledge (IST-1999-10132) Deliverable 9, Aidministrator Nederland b.v., April. See http://www.ontoknowledge.org/.

Broekstra, J. and Kampman, A. (2001b) Sesame: A generic Architecture for Storing and Querying RDF and RDF Schema. Deliverable 10, On-To-Knowledge project, October. http://www.ontoknowledge.org/downl/del10.pdf

Broekstra, J., Fluit, C. and van Harmelen, F. (2000), The state of the art on representation and query languages for semistructured data, IST-199-10132 On-To-Knowledge Project, Deliverable 8. [http://www.ontoknowledge.org/del.shtml]

Broekstra, J., Kampman, A. and van Harmelen, F. (2002) Sesame: an architecture for storing and querying RDF data and schema information, In Fensel, D., Hendler, J., Lieberman, H. and Wahlster, W. (eds.) Semantic Web Technology, Boston, MA: MIT Press, in press.

Buzan, T. (1974) Use Your Head. BBC Books.

Cadoli, M. and Schaerf, M. (1995) Tractable reasoning via approximation, Artificial Intelligence, 74(2): 249–310.

Carlsson, P., Ygge, F. and Andersson, A. (2001) Extending Equilibrium Markets, IEEE Intelligent Systems, 16(4): 18–26.

Carrol, J. and McBride, B. (2001) The Jena Semantic Web Toolkit. Public API, HP-Labs, Bristol. See http://www.hpl.hp.com/semweb/jena-top.html.

Cattel, R.G.G., Barry, D., Berler, M., Eastman, J., Jordan, D., Russell, C., Schadow, O., Stanienda, T. and Velez, F. (2000) The Object Database Standard: ODMG 3.0. Morgan Kaufmann.

Chamberlin, D., Florescu, D., Robie, J., Simeon, J. and Stefanescu, M. (2001) XQuery: A Query Language for XML. Working draft, World Wide Web Consortium, February. See http://www.w3.org/tr/xquery/

Chaudhri, V. K., Farquhar, A., Fikes, R., Karp, P.D. and Rice, J. (1998) OKBC: A programmatic foundation for knowledge base interoperability. In Proceedings of the 15th National Conference on Artificial Intelligence (AAAI'98), pp. 600–607.

Clark, J. (1999) XSL Transformations (XSL-T), W3C Recommendation, 1999. http://www.w3.org/tr/xslt/

Constant, D., Sproull, L. and Kiesler, S. (1996) The kindness of strangers: the usefulness of electronic weak ties for technical advice, Organisation Science, 7(2): 119–135.

Cook, S.D.N. and Seely-Brown, J. (1998) Bridging epistemologies: the generative dance between organizational knowledge and organizational knowing, Organization Science, 10(July–August): 381–400.

Corcho, O. and Gomez-Perez, A. (2000) A roadmap to ontology specification languages, In Dien, R. and Corby, O (eds.), Knowledge Engineering and Knowledge Management; Methods, Models and Tools, Proceedings of the 12th International Conference EKAW, Juan-les-Pins, France, October 2–6 LNCS 1937, pp. 80–96.

Dalal, M. (1996) Semantics of anytime family of reasoners, In Wahlster, W. (ed.), Proceedings of ECAI'96, pp. 360–364.

Das, A., Wu, W., McGuinness, D.L. and Cheyer, A. (2001) Industrial strength ontology management for e-business applications. In the Proceedings of International Semantic Web Working Symposium (SWWS), July 30–August 1, Stanford University, CA.

Davies, N.J. (2000a) Knowledge management, BT Technology Journal, 18, 1, Kluwer, Netherlands.

Davies, N.J. (2000b) Supporting virtual communities of practice. In Roy, R. (ed.), Industrial Knowledge Management, Springer-Verlag.

Davies, N.J. and Chen, C. (1999) Integrating spatial, semantic, and social structures for knowledge management. In Proceedings HICSS-99, Hawaii, January.

Decker, S., Erdmann, M., Fensel, D. and Studer, R. (1999) Ontobroker: ontology based access to distributed and semi-structured information. In Meersman, R., Tari, Z. and Stevens, S. (eds.), Database Semantics: Semantic Issues in Multimedia Systems. Kluwer Academic.

Declerck, T. and Neumann, G. (2000) Using a parameterizable and domain-adaptive information extraction system for annotating large-scale corpora? In Proceedings of the LREC'2000 Workshop Information Extraction meets Corpus Linguistics, Athens.

Deiters, W., Lucas, R. and Weber, T. (2000) Skill-Management: ein Baustein für das Management flexibler Teams. Information Management and Consulting, 15(3): 54–60.

Dimitrov, M. (2000) XML standards for ontology exchange. In Proceedings of Onto-Lex 2000: Ontologies and Lexical Knowledge Bases, September 8–10, Sozopol.

Ding, Y. and Fensel, D. (2001) Ontology library systems: The key for successful ontology reuse. The First Semantic Web Working Symposium (SWWS1), July 29–August 1, Stanford, CA.

Ding, Y. and Fensel, D. (2002) Ontology Library Systems: The essential ontology management issue for the Semantic Web, Knowledge Engineering Review, submitted.

Ding, Y., Fensel, D., Klein, M. and Omelayenko, B. (2001) Ontology management:-survey, requirements and directions. Deliverable 4, On-To-Knowledge project, June 2001. http://www.ontoknowledge.org/downl/del4.pdf

Domingue, J. (1998) Tadzebao and webonto: discussing, browsing, and editing ontologies on the web. In Proceedings of the 11th Knowledge Acquisition for Knowledge-Based Systems Workshop, April 18–23, Banff, Canada.

Domingue, J. and Motta, E. (1999) A knowledge-based news server supporting ontology-driven story enrichment and knowledge retrieval. In Fensel, D. and Studer, R. (eds.), Proceedings of the 11th European Workshop on Knowledge Acquisition, Modeling, and Management (EKAW '99), LNAI 1621, Springer-Verlag.

Driver, M., Brousseau, K.R. and Hunsaker, P.L. (1993) The Dynamic Decision Maker. San Francisco, CA: Jossey-Bass.

Duineveld, A.J., Stoter, R., Weiden, M.R., Kenepa, B.and Benjamins, V.R. (1999) Won-derTools? A comparative study of ontological engineering tools. In Proceedings of the Twelfth Workshop on Knowledge Acquisition, Modeling and Management. October 16–21, Banff, Canada.

Düsterhöft, A., Heuer, A., Klettke, M. and Priebe, D. (1999) GETESS: Text-orientierter Anfrage- und Suchdienst im Internet. Grundlagen von Datenbanken 1999, pp. 17–21.

Engels, R.H.P. and Bremdal, B.A. (2000) Information extraction: state-of-the-art report. Deliverable 5 of the EU 5th Framework Project OntoKnowledge (IST-1999-10132).

Engels, R. and Bremdal, B. (2001) Ontology Extraction Tool. On-To-Knowledge deliverable D-6, CognIT a.s, 2001.

Farquhar, A., Fickas, R. and Rice, J. (1996) The Ontolingua Server: A tool for collaborative ontology construction. In Proceedings of the 10th Banff Knowledge Acquisition for Knowledge Based System Workshop (KAW'95), November, Banff, Canada.

Fensel, D. (2001) Ontologies: Silver Bullet for Knowledge Management and Electronic Commerce. Berlin: Springer-Verlag.

Fensel, D. and van Harmelen, F. (2000) Project Presentation On-To-Knowledge: Content-driven Knowledge-Management Tools through Evolving Ontologies Deliverable 33 of the EU 5th framework project OntoKnowledge (IST-1999-10132).

Fensel, D., van Harmelen, F., Klein, M. and Akkermans, H. (1999) OntoKnowledge: Ontology Based Tools for Knowledge Management. In: Proceedings of the business and Ecommerce Conference, Madrid, Spain.

Fensel, D., Horrocks, I., van Harmelen, F., McGuinness, D. and Patel-Schneider, P.F. (2001) OIL: ontotlogy infrastrucutre to enable the semantic web, IEEE Intelligent System, 16(2).

Fensel, D., Hendler, J., Lieberman, H. and Wahlster, W. (eds.) (2002a) Semantic Web Technology, Boston, MA: MIT Press.

Fensel, D., Hendler, J., Lieberman, H. and Wahlster, W. (2002b) Introduction, In Fensel, D., Hendler, J., Lieberman, H. and Wahlster, W. (eds.), Semantic Web Technology, Boston, MA: MIT Press.

Fensel, D., van Harmelen, F., Akkermans, H., Klein, M., Broekstra, J., Fluyt, C., van der Meer, J., Schnurr, H.-P., Studer, R., Davies, J., Hughes, J., Krohn, U., Engels, R., Bremdahl, B., Ygge, F., Reimer, U. and Horrocks, I. (2000c) OnToKnowledge: ontology-based tools for knowledge management. In Proceedings of the eBusiness and eWork 2000 (EMMSEC 2000) Conference, October, Madrid, Spain.

Fensel, D., Horrocks, I., van Harmelen, F., Decker, S., Erdmann, M. and Klein, M. (2000d) OIL in a nutshell. In Dieng, R. and Corby, O. (eds.), Knowledge Acquisition, Modeling and Management: Proceedings of the European Knowledge Acquisition Conference (EKAW-2000), Berlin: Springer.

Fikes, R. and McGuinness, D.L. (2001) An axiomatic semantics for rdf, rdf schema, and daml+oil. Stanford University KSL Technical Report KSL-01-01. http://www.ksl.stanford.edu/people/dlm/daml-semantics/abstract-axiomatic-semantics.html.

Franconi, E., Grandi, F. and Mandreoli, F. (2000a) Schema evolution and versioning: a logical and computational characterization. In Database Schema Evolution And Meta-Modeling - Ninth International Workshop on Foundations of Models and Languages for Data and Objects, Schloss Dagstuhl LNCS no. 2065, pp. 85–99, Germany, September 18–21.

Franconi, E., Grandi, F. and Mandreoli, F. (2000b) A semantic approach for schema evolution and versioning of OODB. In Proceedings of the 2000 International Workshop on Description Logics (DL2000), Aachen, Germany, August 17–19, pp. 99–112.

Fridman N., Fergerson, R. and Musen, M. (2000) The knowledge model of Protégé-2000: Combining interoperability and flexibility. In Proceedings of European Knowledge Acquisition Workshop (EKAW) 2000, LNCS 1937, pp. 17–32. Springer.

Genesereth, M.R. and Fikes, R.E. (1992) Knowledge Interchange Format, Version 3.0 Reference Manual. Technical Report Logic-92-1, Stanford University.

Goldman-Segall, R. and Rao, S.V. (1998) A collaborative online digital data tool for creating living narratives. In Organisational Knowledge Systems, 31st Hawaii International Conference on Systems Science, Hawaii, USA.

Gomez-Perez, A. (1996) A framework to verify knowledge sharing technology. Expert Systems with Applications, 11(4): 519–529.

Granovetter, M. (1974) The strength of weak ties, American Journal of Sociology, 78: 1360–1380.

Granovetter, M. (1982) The strength of weak ties: a network theory revisited, In, Marsden, P. and Nan, L. (Eds.), Social Structure and Network Analysis, California: Sage Publications.

Grefenstette, G. and Tapanainen, P. (1994) What is a Word, What is Sentence? Problems of Tokenization, Grenoble: Rank Xerox Research Centre.

Grosso, W.E., Eriksson, H., Fergerson, R.W., Gennari, J.H., Tu, S.W. and Musen, M.A. (1999) Knowledge modelling at the millenium (the design and evolution of protégé-2000. In Proceedings of Knowledge Acquisition Workshop (KAW-99).

Gruber, T.R. (1993) A translation approach to portable ontology specifications. Knowledge Acquisition, 5(2).

Guarino, N. and Welty, C. (2000) identity, unity, and individuality: towards a formal toolkit for ontological analysis. Proceedings of ECAI-2000, August.

Gustavsson, R. (1999) Agents with power, Communications of the ACM, 42(3): 41–47.

Haarslev, V., Mueller, R. and Turhan, A.-Y. (2001) Exploiting Pseudo Models for Tbox and Abox Reasoning in Expressive Description Logics, IJCAI2001.

Handschuh, S. (2001) Ontoplugins – a flexible component framework. Technical Report, University of Karlsruhe, May.

Handschuh, S. and Staab, S. (2001) Authoring and Annotation of Web Pages in CREAM, submitted.

Handschuh, S., Staab, S. and Mädche, A. (2001) CREAM – Creating relational metadata with a component-based, ontology-driven annotation framework. In K-CAP 2001 – Proceedings of the First International ACM Conference on Knowledge Capture. October 21–23, Victoria, BC, Canada.

Hansen, M.T. (1997) The Search-Transfer Problem: The Role of Weak Ties in Sharing Knowledge Across Organisation Subunits, Working Paper, Harvard Business School.

Harman, D. (1992) Ranking algorithms. In Frakes, W. and Baeza-Yates, R. (eds.), Information Retrieval, Englewood Cliffs, NJ: Pentice-Hall.

Hayes, P. (2001) RDF Model Theory. W3C Working Draft. http://www.w3.org/tr/rdf-mt/

Hearst, M.A. (1998) Information integration. IEEE Intelligent Systems, September/ October: 12–24.

Hegel, G. (1812) Wissenschaft der Logik, Nuremberg, Germany. English translation available at http://www.marxists.org/reference/archive/hegel/works/hl/

Heijst, G., van Schreiber, A.T. and Wielinga, B.J. (1997) Using explicit ontologies in KBS development. International Journal of Human-Computer Studies, 45: 183–292.

Hendler, J. (2000) DARPA Agent Markup Language. Technical Report, Defense Advanced Research Projects Agency, http://www.daml.org/.

Hollunder, B. and Baader, F. (1991) Qualifying number restrictions in concept languages. In Proceedings of the 2nd International Conference on the Principles of Knowledge Representation and Reasoning (KR'91), pp. 335–346.

Horrocks, I. (1998) Using an expressive description logic: FaCT or fiction? In Proceedings of KR 1998, pp. 636–649. Morgan Kaufmann.

Horrocks I. and Sattler U. (2001) Ontology Reasoning in the SHOQ(D) Description Logic, IJCAI2001.

Horrocks, I. and Tessaris, S. (2000) A conjunctive query language for description logic Aboxes. In Proc. of AAAI 2000, pp. 399–404.

Horrocks, I., van Harmelen, F., Patel-Schneider, P., Berners-Lee, T., Brickley, D., Connoly, D., Dean, M., Decker, S., Fensel, D., Hayes, P., Hein, J., Hendler, J., Lassila, O., McGuinness, D. and Stein, L.A. (2001a) DAML+OIL. http://www.daml.org/2001/03/daml+oil-index.html, March.

Horrocks, I., Fensel, D., Broekstra, J., Decker, S., Erdmann, M., Goble, C., van Harmelen, F., Klein, M., Staab, S., Studer, R. and Motta, E. (2001b) OIL: The Ontology Inference Layer. Technical Report IR-479, Faculty of Sciences, Vrije Universiteit Amsterdam, September.

Hotho, A., Mädche, A., Staab, S. and Studer, R. (2001) SEAL-II - The soft spot between richly structured and unstructured knowledge. Journal of Universal Computer Science, 7(7): 566–590.

Hu, J., Kashi, R., Lopresti, D. and Wilfon, G. (1999) Table detection across mutiple media. In Proceedings of the Document Layout Interpretation and its Applications (DLIA99) Workshop, Bangalore, India.

Iosif, V. and Ygge, F. (2001) EnerSearch Virtual Organisation Case Study: Requirements Analysis Document, On-To-Knowledge EU-IST-1999-10132 Project Deliverable D27, January, Malmö, Sweden: EnerSearch AB.
http://www.ontoknowledge.org/ and http://www.enersearch.se/

Iosif, V. and Ygge, F. (2002) EnerSearch Virtual Organisation Case Study: VE Prototype, On-To-Knowledge EU-IST-1999-10132 Project Deliverable D28, January, Malmö, Sweden: EnerSearch AB.
http://www.ontoknowledge.org/ and http://www.enersearch.se/

ISO (1999) Information Technology-Database Language SQL. Standard no. ISO/IEC 9075:1999, International Organization for Standardization. Available from American National Standards Institute, New York.

Jin, Y., Decker, S. and Wiederhold, G: (2001) OntoWebber: Model-Driven Ontology-Based Web Site Management. In Proceedings of the 1st International Semantic Web Working Symposium (SWWS '01), July 29–August 1, Stanford, CA: Stanford University.

Kaplan, R. and Bresnan, J. (1982) Lexical-functional grammar: A formal system for grammatical representation. In Bresnan, J. (ed.), The Mental Representation of Grammatical Relations, Cambridge, MA: MIT Press.

Karp, P.D., Chaudhri, V.K. and Thomere, J. (1999) XOL: An XML-Based Ontology

Exchange Language, Version 0.3, July 3, 1999.
ftp://smi.stanford.edu/pub/bio-ontology/ontologyexchange.doc

Karvounarakis, G., Christophides, V., Plexousakis, D. and Alexaki, S. (2000) Querying community web portals. Technical Report, Heraklion, Greece: Institute of Computer Science, FORTH. See http://www.ics.forth.gr/proj/isst/rdf/rql/rql.pdf.

Kieninge, T. and Denge, A. (1999) The T-RECS approach for table structure recognition and table border determination. In Proceedings of the Document Layout Interpretation and its Applications (DLIA99) Workshop, Bangalore, India.

Kiryakov, A., Simov, K.Iv. and Dimitrov, M. (2001) OntoMap - the guide to the upper-level. In Proceedings of the International Semantic Web Working Symposium (SWWS), July 30–August 1, Stanford University, CA.

Kiryakov, A., Simov, K.Iv. and Ognyanov, D. (2002) OntologyMiddleware Module: Analysis and Design. Deliverable 38, On-To-Knowledge Project, February. http://www.ontoknowledge.org/downl/del38.pdf

Kitcharoensakkul, S. and Wuwongse, V. (2001) Towards a unified version model using the resource description framework (RDF). International Journal of Software Engineering and Knowledge Engineering, 11(6).

Klabunde, R., Carstensen, K.-U., Ebert, C., Endriss, C., Jekat, S. and Langer, H. (2001) Informationsextraktion. In Klabunde, R., Carstensen, K.-U., Ebert, C., Endriss, C., Jekat, S. and Langer, H. (eds.), Computerlinguistik and Sprachtechnologie - Eine Einfuhrung. Heidelberg: Spektrum Akademischer.

Klein, M. (2001) Combining and relating ontologies: an analysis of problems and solutions. In Gomez-Perez, A., Gruninger, M., Stuckenschmidt, H. and Uschold, M. (eds.), Workshop on Ontologies and Information Sharing, IJCAI'01, August 4–5, Seattle, WA.

Klein, M. and Fensel, D. (2001) Ontology versioning for the semantic web. In Proceedings of the International Semantic Web Working Symposium (SWWS), July 30–August 1, Stanford University, CA.

Klemke, R. and Nick, A. (2001) Case studies in developing contextualising information systems. In Proceedings of CONTEXT 2001, LNCS 2116, pp. 457–460, Springer.

Knowledge Management Group at AIFB (2001) Engineering Environment OntoEdit. Technical Report, Angewandte Informatik und Formale Beschreibungs-verfahren, University of Karlsruhe. http://ontoserver.aifb.uni-karlsruhe.de

Larsson, R., Driver, M. and Sweet, P. (1998) Interactive marketing and organisation learning dynamics: implications of ISES research for energy companies. In Ottosson, H., Akkermans, J.M. and Ygge, F. (eds.), Information Society Energy System – The ISES Project 1996–1998, pp. 31–60. Malmö, Sweden: EnerSearch AB. Also available from: http://www.enersearch.se

Larsson, R., Schönknecht, J., Sweet, P. and Driver, M. (1999) The customer side of energy saving activities: exploring attitudes and interests on the Karlshamn's energy market. In Akkermans, J.M. and Ottosson, H. (eds.), Energy Efficiency In A Deregulated Market – The KEES Project, Chapter 3, pp. 39–60. Malmö, Sweden: EnerSearch AB. Also available from: http://www.enersearch.se

Lassila, O. and Swick, R.R. (eds.) (1999) W3C; Resource Description Framework

(RDF) Model and Syntax Specification.
http://www.w3.org/tr/1999/rec-rdf-syntax-19990222/

Lau, T. and Sure, Y. (2002) Introducing ontology-based skills management at a large insurance company. In Proceedings of the Modellierung 2002, March, Tutzing, Germany.

LeBourgeois, F. and Souafi-Bensafi, S. (1999) RASADE: Automatic recognition of structured document using typography and spatial inference. In Proceedings of the Document Layout Interpretation and its Applications (DLIA99) Workshop, Bangalore, India.

Lebowitz, M. (1983): Memory-based parsing. Artificial Intelligence, 21.

Lee, D. and Chu, W. (2001) CPI: constraint-preserving inlining algorithm for mapping XML DTD to relational schema, Data and Knowledge Engineering, 39: 3–25.

Lerner, B.S. (2000) A model for compound type changes encountered in schema evolution. ACM Transactions on Database Systems, 25(1): 83–127.

Liao, M., Hinkelmann, K., Abecker, A. and Sintek, M. (1999) A competence knowledge base system for the organisational memory. In Puppe, F. (ed.), XPS-99: Knowledge-Based Systems, LNAI 1570, pp. 125–137.

López, F. (1999) Overview of methodologies for building ontologies. In Proceedings of the IJCAI-99 Workshop on Ontologies and Problem-SolvingMethods: Lessons Learned and Future Trends. CEUR Publications.

Lopez, M.F., Gomez-Perez, A., Sierra, J.P. and Sierra, A.P. (1999) Building a chemical ontology using methontology and the ontology design environment. Intelligent Systems, 14(1).

Mädche, A. and Staab, S. (2000) Discovering conceptual relations from text. In Proceedings of the 14th European Conference on Artificial Intelligence (ECAI 2000), Berlin: IOS Press.

Mädche, A. and Staab, S. (2001) Ontology learning for the semantic web. IEEE Intelligent Systems, 16(2): 72–79.

Mädche, A., Neumann, G. and Staab, S. (1999) A generic architectural framework for text knowledge acquisition. Technical Report, Institute AIFB.

Mädche, A., Nack, F., Santini, S., Staab, S. and Steels, L. (2000) Emergent semantics. IEEE Intelligent Systems, Trends and Controversies, 17(2), 2002.

Mädche, A., Staab, S., Studer, R., Sure, Y. and Volz, R. (2002) SEAL – Tying up information integration and web site management by ontologies. IEEE-CS Data Engineering Bulletin (Special Issue on Organizing and Discovering the Semantic Web), March, in press.

Manber, U., Smith, M. and Gopal, B. (1997) WebGlimpse – Combining Searching and Browsing. Usenix 97 Technical Conference.

Marchionini, G. (1995) Information seeking in electronic environments. Cambridge Series on Human-Computer Interaction. Cambridge University Press.

Mattox, D., Maybury, M. and Morey, D. (1999) Enterprise expert and knowledge discovery. In Proceedings of the International Conference on Human Computer Interaction, pp. 303–307.

Maurer, H. and Sapper, M. (2001) E-Learning has to be seen as part of general knowledge management. In Proceedings of ED-MEDIA 2001 World Conference on

Educational Multimedia, Hypermedia and Telecommunications, pp. 1249–1253, Charlottesville, VA: Tampere, CE.

Maxwell, C. (2000) The future of work – understanding the role of technology, BT Technology Journal, 18: 1.

Maybury, M., D'Amore, R. and House, D. (2000) Automatic expert finding. International Journal of Technology Research Management, 43(6): 12–15

Mello, R. and Heuser, C. (2001) A rule-based conversion of a DTD to a conceptual schema, In Kunii, H., Jojodia, S. and Solvberg, A. (eds.), Conceptual Modeling - ER'2001, LNCS 2224, pp. 133-148, Yokohama, Japan, November 27–30, Springer.

McDonald, D. and Ackerman, M. S. (1998) Just talk to me: a field study of expertise location. In Proceedings of the ACM Confrerence. on Computer Supported Cooperative Work (CSCW'98).

McGuinness, D.L. (2000) Conceptual modelling for distributed ontology environment. In Proceedings of the Eighth International Conference on Conceptual Structures Logical, Linguistic, and Computational Issues (ICCS2000), Darmstadt, Germany, August 14–18.

McGuinness,. D.L., Fikes, R., Rice, J. and Wilder, S. (2000) An environment for merging and testing large ontologies. In Proceedings of the Seventh International Conference on Principles of Knowledge Representation and Reasoning (KR2000). Breckenridge, CO, April 12–15.

Miller, L. (2001) RDF squish query language and Java implementation. Public Draft, Institute for Learning and Research Technology.
See http://ilrt.org/discovery/2001/02/squish/.

Mitra, P., Wiederhold, G. and Kersten, M. (2000) A graph-oriented model for articulation of ontology interdependencies. In: Zaniolo, C., Lockemann, P., Scholl, M. and Grust, T. (eds.), Advances in Database Technology - EDBT 2000, 7th International Conference on Extending Database Technology, LNCS 1777, pp. 86–100, Konstanz, Germany, March 27–31, Springer-Verlag.

Mladenic, D. (1999) Text-learning and related intelligent agents: a survey. IEEE Intelligent Systems 14(4): 44–54.

Nejdl, W., Wolf, B., Staab, S. and Tane, J. (2002) EDUTELLA: Searching and Annotating Resources within an RDF-based P2P Network. In 1st International Workshop on Peer-to-Peer Systems (IPTPS02), submitted.

Nonaka, I. (1994) A dynamic theory of organisational knowledge creation. Organisation Science, 5: 1.

Noy, F.N. and Hafner, C.D. (1997) The state of the art in ontology design: a survey and comparative review. AI Magazine 4: 53–74.

O'Leary, D.E. and Studer, R. (eds.) (2001) Knowledge management, IEEE Intelligent Systems, 16(1).

Omelayenko, B. and Fensel, D. (2002) Scalable document integration for B2B electronic commerce, Electronic Commerce Research Journal, submitted.

OntoEdit, http://ontoserver.aifb.uni-karlsruhe.de/ontoedit/

Oram, A. (ed.) (2001) Peer-to-peer: harnessing the benefits of a disruptive technology, Sebastobol: O'Reilly.

Palmer, D.D. and Hearst, M.A. (1994) Adaptive sentence boundary disambiguation. In

Proceedings of the Fourth ACL Conference on Applied Natural Language Processing. Stuttgart.

Pan, J. and Horrocks, I. (2001) Metamodeling architecture of web ontology languages. In the Proceedings of the International Semantic Web Working Symposium (SWWS), July 30–August 1, Stanford University, CA.

Patel-Schneider, P. (ed.) (2001) A Model-Theoretic Semantics for DAML+OIL, March.
http://www.daml.org/2000/12/model-theoretic-semantics.html, November.

Patel-Schneider, P. and van Harmelen, F. (2001) Coordination points between RDF(S) and DAML+OIL.
http://www.daml.org/2001/07/rdfs-daml+oil-coordination.html

Pinto, H.S., Gómez-Pérez, A. and Martins, J.P. (1999) Some issues on ontology integration. In Proceedings of the Workshop on Ontologies and Problem Solving Methods IJCAI-99, Stockholm, Sweden.

Pollard, C. and Sag, I. (1994) Head-driven Phrase Structure Grammar. Chicago, IL: University of Chicago Press.

Polyani, M. (1966) The Tacit Dimension. London: Routledge and Paul.

Ram, A. and Moorman, K. (1999) Introduction: toward a theory of reading and understanding. In Ram, A. and Moorman, K. (eds.) Understanding Language Understanding: Computational Models of Reading. Cambridge, MA: MIT Press.

Roddick, J.F. (1995) A survey of schema versioning issues for database systems. Information and Software Technology, 37(7): 383–393.

Salton, G. (1989) Automatic Text Processing. Reading, MA: Addison-Wesley.

Schreiber, G., Akkermans, H., Anjewierden, A., de Hoog, R., Shadbolt, N., Van de Velde, W. and Wielinga, B. (1999) Knowledge engineering and management - the CommonKADS Methodology. Cambridge, MA: The MIT Press.

Seely-Brown, J. and Duguid, P. (1991), Organisational learning and communities of practice, Organisational Science, 2(1).

Shank, R.C. (1975) Conceptual Information Processing. Amsterdam: North Holland.

Shipman, F.M., Marshall, C.C. and Moran, T.P. (1995) Finding and using implicit structure in human-organized spatial layouts of information. In CHI-95.

Shneiderman, B. (1996) The eyes have it: a task by data type taxonomy of information visualizations, In the Proceedings of the IEEE Symposium on Visual Languages '96, pp. 336–343, September, Los Alamitos, CA, IEEE.

Sintek, M. and Decker, S. (2001) TRIPLE - An RDF query, inference, and transformation language, In Proceedings of the Workshop on Deductive Databases and Knowledge Management (DDLP-2001), Tokio, Japan, October 20–22.

Spitz, L. (1999) Style-directed document recognition. In Proceedings of the Document Layout Interpretation and its Applications (DLIA99) Workshop, Bangalore, India.

Staab, S. and Mädche, A. (2001) Knowledge portals - ontologies at work, AI Magazine, 21(2): 63–75.

Staab, S. and Schnurr, H.-P. (2000) smart task support through proactive access to organizational memory, Knowledge-Based Systems, 3(5): 251–260.

Staab, S., Schnurr, H.-P., Studer, R. and Sure, Y. (2001) Knowledge processes and ontologies, IEEE Intelligent Systems, 16(1): 26–35.

Stojanovic, L., Staab, S. and Studer, R. (2001) eLearning based on the semantic web. In Proceedings of the WebNet 2001 - World Conference on the WWW and the Internet, October 23–27, Orlando, FL.

Sure, Y., Maedche, A. and Staab, S. (2000) Leveraging corporate skill knowledge - from ProPer to OntoProper. In Reimer, U. (ed.), 3rd International Conference on Practical Aspects of Knowledge Management PAKM 2000.

Sure, Y., Staab, S., Angele, J., Wenke, D. and Maedche, A. (2002) OntoEdit: Guiding ontology development by methodology and inferencing. In Prestigious Applications of Intelligent Systems (PAIS), in conjunction with ECAI 2002, July 1–26, Lyon, France, submitted.

Swartout, B., Patil, R., Knight, K. and Russ, T. (1996) Toward distributed use of large-scale ontologies. In Proceedings of the 10th Knowledge Acquisition Workshop (KAW'96), November, Banff, Canada.

Sweet, P. (2001) Designing interactive value development: perspectives and strategies for high-precision marketing, PhD Thesis, Lund University, May, Malmö, Sweden: EnerSearch AB.

Sweet, P., Olsson, M. and Akkermans, J.M. (2000) PLC Service Business Model Development and Market Survey Instrument, PALAS EU-IST-1999-11379 Project Powerline as an Alternative Local AccesS, Deliverable D7, Malmö, Sweden:, EnerSearch AB, December.

Also available from http://palas.regiocom.net and http://www.enersearch.se

Taubner, D. and Brössler, P. (2000) The people make the project. Project control: the human factor. In Proceedings of ESCOM-SCOPE 2000, pp. 105–113.

Tennison, J. and Shadbolt, N. (1998) APECKS: a tool to support living ontologies. In Proceedings of the 11th Knowledge Acquisition Workshop (KAW'98), April, Banff, Canada.

The Gene Ontology Consortium (2000) Gene ontology: tool for the unification of biology. Nature Genetics, 25(1): 25–29.

Tiwana, A. (2000) The Knowledge Management Toolkit. Upper Saddle River, NJ: Prentice Hall PTR.

Uschold, M. and Grueninger, M. (1996) Ontologies: principles, methods and applications. Knowledge Sharing and Review, 11(2).

Uschold, M. and King, M. (1995) Towards a methodology for building ontologies. In Workshop on Basic Ontological Issues in Knowledge Sharing, held in conjunction with IJCAI-95, Montreal, Canada.

Uschold, M., Healy, M., Williamson, K., Clark, P. and Woods, S. (1998) Ontology reuse and application. In Guarino, N. (ed.), Formal Ontology in Information Systems (FOIS'98), Treno, Italy, June 6–8, Amsterdam: IOS Press.

van Harmelen, F., Patel-Schneider, P.F. and Horrocks, I. (eds.) (2001a) Reference description of the DAML+OIL (March) ontology markup language. http://www.daml.org/2001/03/reference.html

van Harmelen, F., Kampman, A. and Broekstra, J. (2001b) Interoperability and scalability of on-to-knowledge tools. On-To-Knowledge deliverable D-X1, Vrije Universiteit Amsterdam.

Ventrone, V. and Heiler, S. (1991) Semantic heterogeneity as a result of domain evolution. SIGMOD Record (ACM Special Interest Group on Management of Data), 20(4): 16–20.

Visser, P.R.S. and Bench-Capon, T.J.M. (1998) A comparison of four ontologies for the design of legal knowledge systems. Artificial Intelligence and Law, 6: 27–57.

Visser, P.R.S., van Kralingen, R.W. and Bench-Capon, T.J.M. (1997a) A method for the development of legal knowledge systems. In Proceedings of the Sixth International Conference on Artificial Intelligence and Law (ICAIL'97), Melbourne, Australia.

Visser, P.R.S., Jones, D.M., Bench-Capon, T.J.M. and Shave, M.J.R. (1997b) An analysis of ontological mismatches: heterogeneity versus interoperability. In AAAI 1997 Spring Symposium on Ontological Engineering, Stanford, CA.

Wasserman, R. (2000) Resource-bounded belief revision, Ph.D. thesis, ILLC, Faculty of Science, University of Amsterdam.

Ygge, F. and Akkermans, J.M. (1999) Decentralized markets versus central control - a comparative study, Journal of Artificial Intelligence Research, 11: 301–333. Also available from: http://www.jair.org

Younker, E. (1998) Skills management: the linchpin for IT work force management. InSide GartnerGroup, 2–4.

Index

Printed in the United Kingdom by
Lightning Source UK Ltd., Milton Keynes
137329UK00001B/88/P